Housing and labour markets

Housing and labour markets

Building the connections

edited by

JOHN ALLEN & CHRIS HAMNETT

London
UNWIN HYMAN
Boston Sydney Wellington

Published by the Academic Division of
Unwin Hyman Ltd
15/17 Broadwick Street, London W1V 1FP, UK

Unwin Hyman Inc.
955 Massachusetts Avenue, Cambridge, MA 02139, USA

Allen & Unwin (Australia) Ltd
8 Napier Street, North Sydney, NSW 2060, Australia

Allen & Unwin (New Zealand) Ltd
in association with the Port Nicholson Press Ltd
Compusales Building, 75 Ghuznee Street, Wellington 1, New Zealand

First published in 1991

British Library Cataloguing in Publication Data
Housing and labour markets: building the connections.
1. Great Britain. Housing market. Related to labour market
2. Great Britain. Labour market. Related to housing market
I. Allen, John II. Hamnett, Chris
363.50941

ISBN 0–04–445443–0

Library of Congress Cataloging in Publication Data
Housing and labour markets: building the connections/edited by John Allen and Chris Hamnett.
p. cm.
Includes bibliographical references and index.
ISBN (invalid) 0–00–445443–0
1. Housing–Great Britain. 2. Labor market–Great Britain.
I. Allen, John, 1951– . II. Hamnett, Chris.
HD7333.A3H68 1990
363.5′1′0941–dc20 90–37405
CIP

Typeset in 10 on 11 point Bembo
by Falcon Typographic Art Ltd
and printed and bound in Great Britain by
Biddles Ltd, Guildford and King's Lynn

Contents

Contributors

John Allen, Lecturer, Faculty of Social Sciences, The Open University.

James Barlow, Research Fellow, Centre for Urban and Regional Research, School of Cultural and Community Studies, University of Sussex.

Tony Champion, Lecturer, Department of Geography, University of Newcastle upon Tyne.

Mike Coombes, Principal Research Associate, Centre for Urban and Regional Studies, University of Newcastle upon Tyne.

Ray Forrest, Research Fellow, School for Advanced Urban Studies, University of Bristol.

Mark Goodwin, Lecturer, Department of Geography, University College, Lampeter.

Chris Hamnett, Senior Lecturer, Faculty of Social Sciences, The Open University.

Ray Hudson, Professor, Department of Geography, University of Durham.

Peter Marcuse, Professor, Department of Urban Planning, Columbia University, USA.

Moira Munroe, Research Fellow, Centre for Housing Research, University of Glasgow.

Alan Murie, Professor, Department of Housing and Planning, Heriot Watt University, Edinburgh.

Bill Randolph, Research Officer, National Federation of Housing Associations, 175 Grays Inn Road, London WC2.

John Salt, Senior Lecturer, Department of Geography, University College London.

Mike Savage, Lecturer, Department of Sociology, University of Keele.

Sophie Watson, Lecturer, Department of Social Policy, University of Bristol.

List of figures

List of tables

PART ONE

Building the connections

Introduction

Homes and jobs – or the lack of them – constitute two of the most central and important areas of social life. It is therefore not surprising that a large volume of academic research has focused on work and housing. What is surprising is that despite the strong links between home and work in people's daily lives, and in the structure of cities and regions, there have been few systematic studies of the relationship between the two areas. Despite the major social and economic problems and policy issues generated by the interaction of housing and labour market changes, the two areas have traditionally been analyzed by economists, sociologists, geographers and others as though they existed in virtual isolation, and the relationships between them have, until recently, either been glossed over or taken for granted.

This tendency to treat homes and work and housing and labour markets in isolation has been reinforced by an academic division of labour. There are numerous economic and sociological studies of the labour market and of labour processes, and there are also many studies of the structure and operation of the housing market and its role in the production of socio-spatial segregation. But few of these studies have linked the two. In economic and industrial geography, research has tended to focus on the geography of production and employment, while in urban social geography, the focus has been on where people live as though the two things are unrelated. Thus, while the importance of the housing market in 'sifting and sorting' the population in space is widely accepted, the occupational and income structure and indeed the very existence of the population has been commonly treated as an unproblematic given. Nor, for the most part, are variations in the structure of the housing stock or the operation of the housing market explicitly problematized. Like the population, the housing stock is simply assumed to be there, awaiting or containing its quota of households.

This collection is an attempt to raise our understanding of the connections between the two markets; between homes and jobs. It sets out to problematize precisely this connection or rather connections. Its aim, however, is not to provide a classification or definitive list of the different ways in which labour and housing markets connect. Indeed, we are not convinced that such a list would be valuable. In so far as the relationships between labour and housing markets are not static or universal, the nature and the form of the relationships are likely to vary considerably through time and over space, as does the structure of the labour and housing markets themselves. As a result, there are a great many possible relationships. The issue, then, is how we should approach this apparent complexity, in particular how we should conceptualize this variety of relations. Above all, what are the principal lines of differentiation and how may they best be theorized? And to probe these kinds of questions, we need to know something of the kinds of social processes that generate the changing structure of labour and housing markets and the links between them. In many ways, therefore, this book is about setting an agenda for research on labour and housing markets. Its tone is tentative and its purpose is to build some of the links that connect the areas of home and work. First, however, we will spell out some of the reasons why we think it is important to know more about the relations that hold between housing and labour markets.

The market connection

Sometimes the most significant issues to focus upon are among the more obvious or seemingly transparent concerns (hence they are commonly overlooked). Our first reason for exploring the links between labour and housing markets is both political and economic, and takes its brief from geography. It is this. It is a truism that the structures of work and residence are intimately and reciprocally interrelated in and across space. Were this not the case, there would be no necessary spatial link between housing and work. People could live almost anywhere and work somewhere quite different. For those in paid employment they are linked by the time and money costs of the journey to work. And this is both an economic issue for worker and employer alike (and the unemployed), and a political issue for national and local government in their attempts to bring together the two sides in the same place.

The result is a complex patchwork of partially overlapping 'local labour market areas' in which 'the great majority of people live and work' (Goddard, 1983). While improvements in transport have enabled people to live progressively further from their paid employment and broken down the tight local labour market areas of the nineteenth century (which have been extensively documented by Steadman-Jones (1971) and others) the links still exist, though on a larger geographical scale than hitherto. That is not to suggest that housing and labour market areas are spatially congruent.

Although it is possible to delineate local labour markets in terms of a given percentage of residents who commute to work in particular employment centres, the spatial extent of any given labour market is not precisely fixed. There are always some workers who will commute long distances. And in some areas, such as the West Midlands, there will often be a considerable overlap between individual labour markets. Several different local labour markets could be nested within one large housing market or vice versa. In addition, the extent of labour and housing markets is socially differentiated. Highly skilled and highly paid workers are likely to have a much larger potential travel to work area than the less skilled and low paid who are more likely to be constrained by the difficulty and cost of commuting. But, such caveats aside, there is an approximate congruence between housing and labour markets which is imposed by the journey to work. This can be most clearly seen in the case of sparsely populated countries like Australia, Canada, New Zealand, the USA or France. The distances between Perth, Sydney, Adelaide, Melbourne or Brisbane or between Vancouver, Calgary, Toronto and Montreal are simply too great to permit any cross-commuting. But in Britain or the Netherlands the degree of overlap and cross-commuting is much greater, although there are some isolated centres such as Aberdeen. This, of course, is a classic case of the effects of rapid local employment shocks on the local housing market (MacLennan and Jones, 1986). In the USA, several oil dependent cities like Houston have suffered from the turn down in the oil market in the mid-1980s. Unemployment has risen rapidly and house prices have dipped sharply.

The existence of this spatial link between housing and labour markets is of considerable importance in that it provides a framework for understanding many of the important issues and problems which affect individuals and localities, as well as economic and social policy. First, the existence of spatially differentiated labour and housing markets constrains the structure of housing and labour market opportunities open to people in different areas. The situation facing workers and their households in booming or depressed labour markets, easy access or high cost and limited supply markets, is very different. Second, the existence of spatially differentiated and fixed pools of labour poses a variety of opportunities for industrial capital seeking to locate, or change the existing location of production. Third, there is the problem of differential change and mismatches in labour and housing market opportunities. Because houses are immobile and employment is relatively mobile, there is a constant problem of differential adjustment between the size and structure of labour and housing markets in different areas. Jobs move, but people stay, as Booth (1902) commented of nineteenth-century London. Fourth, there is the problem of mobility between different housing and labour markets. While individuals can change either their job or their house within a given labour and housing market without necessarily having to change both, a move from one market to another, usually entails changing both their job and their house. And there can be problems of differential adjustment within the same household if more than one member is in paid employment. As these issues are central to this collection, we will discuss them in turn.

The existence of differential opportunities for employment and housing within local labour and housing markets is of crucial importance for individuals and the communities of which they are a part. Because of the existence of a spatial division of labour at the national, regional and local levels, different countries and different areas have markedly different industrial and occupational structures which are compounded by their different historical legacies. This is most clearly shown in Britain and in other older industrial countries, especially between those regions which had a key role in the development of heavy manufacturing industry in the nineteenth or early twentieth centuries and those regions which have played little or no part in this process, and which have either remained untouched by industrialization or have experienced recent growth of service sector or high-tech industry. In the United States, the division between the rapidly growing sunbelt regions of the South and West and declining snowbelt regions of the North and East has been of considerable importance over the past thirty years. Similar divisions can be found in France, Italy and Germany. There are also important differences between cities and regions, especially between those with a high proportion of the paid workforce in financial services, or research and development, and those with high proportions in assembly work. These differences will be reflected in levels of (un)employment, skill levels, vacancy rates, labour market participation figures and in earned incomes. In buoyant sunbelt labour markets, labour demand and incomes are likely to be much higher than in declining heavy industrial areas or in peripheral cheap labour reserves. It makes a difference to employment opportunities where people live.

Similarly, there are important differences between areas whose housing market contains a high proportion of council housing or cheap older terraced houses, and those with a large or expensive owner-occupied sector where accessibility will be very difficult for those not on high salaries or in multiple earner households. While there is no one to one correspondence between old, declining industrial areas and cheap housing, or between growing industrial areas and high house prices and a small council sector, there is a high degree of correspondence. And to some extent housing costs and incomes are matched. But, for those on low incomes in high cost housing areas there are considerable problems.

These differences between areas are also important for employers. In isolated labour markets where there are only one or two large employers, they can dominate the market and play a key role in determining eligibility criteria and wage levels. These are monopolist labour markets. Arguably, the Norwich Union and HMSO play such a role in the Norwich labour market, certainly for white collar and clerical staff. Similar conditions prevailed in the past in a large number of single industry towns although this dominance has now broken down to a large extent. The conditions in these labour markets may be very different from those where there is no one large employer. In areas which have no tradition of industrial work or which have experienced deindustrialization, there may be pools of cheap labour available for new employers to search out and relocate production. This is now quite commonplace in some types of routine assembly or production and in financial services where data input and processing can be moved to Ireland or the Carribean or anywhere labour is cheap. At the other end of the spectrum, there are sunbelt areas such as Berkshire where the demand for skilled labour is intense and supply is limited, where workers are able to move easily from one employer to another in search of better conditions and higher wages. Such areas are relatively rare.

In general, mobile national or multinational firms possess considerable advantages in that they can use the existence of the friction of distance and spatially fixed local housing markets to move between areas with relative ease in search of pools of cheap or flexible labour. And, in the event of labour difficulties, one area can be played off against another by the threat of closing down production and moving elsewhere. Ford have been particularly successful in this strategy on a national basis, pitting plants in Germany, Belgium or Britain against one another. And even if workers were as mobile, there are considerable family and social constraints on movement between different labour market areas or different countries.

The problems of a single industry town are particularly difficult where the industry declines or closes down. There are examples of this in the declining heavy industrial regions, and in Britain, Corby and Consett have both suffered from the closure of large steel plants which have pushed up male unemployment levels to 30 per cent or more. Similar examples are all too common in the North-eastern United States and some of the old industrial areas of Europe. The key problem of such areas is that the labour force is effectively trapped and deskilled. Locked into a static or declining housing market, they may be unable to move in search of work even if they are willing to do so. Such towns are, of course, one end of a spectrum of employment opportunities and constraints and many of the large industrial cities of the North of Britain have experienced rapid employment decline and high levels of unemployment. In such cases, the only option for many may be to move in search of work to one of the prosperous southern labour markets. Yet, not only may there be a mismatch of skills, the housing market opportunities are simply not available. As a result, Britain has seen a considerable growth of weekly commuting from north to south from the early 1980s by workers who simply cannot gain access to the housing market in the south. Sharing rooms in cheap lodging houses and hotels, they return home only at weekends. The long distance bus stations in Central London are full of such workers on Friday nights.

The problems are similar in many of the prosperous southern areas where labour demand is high, and housing is expensive and in short supply, and workers are forced to live miles away from their place of work and thus commute. In such areas, the supply of housing lags behind employment growth and the adjustment problem is rather the reverse of that in the declining industrial regions where there are plenty of houses and few jobs. This is a classic example of uneven regional development, where inter-regional migration between labour markets is hindered by the shortage of council or rented housing in the south and by the very large regional differences in house prices between north and south. The housing market can compound the employment problems experienced in declining regions while in prosperous regions many workers have been able to earn high salaries and experience rapid capital appreciation on their houses. For those entering the labour market, however, the problem of access to affordable housing can be severe.

These issues have formed a major policy problem for governments. While previous Labour governments tried, not very successfully, to operate a regional policy of taking work to the workers, this has now been largely abandoned in favour of the operation of market forces. Although the Conservative government has offered considerable inducements to employers willing to invest in the depressed regions and has forced some southern counties to increase the projected amount of housing in their strategic plans, the current Conservative administration takes the neo-classical view that, left to itself, the market will reduce regional disparities in the long term – as industries and employers move north to take advantage of cheaper labour and land costs and workers move south for work. Unfortunately, as we have indicated, the housing shortages in the south are such that inter-regional labour migration is severely curtailed. The Conservative government's response to this is to relax planning controls on development in the south and to attempt to dismantle the large council housing monopolies. But, while the level of migration from the South-east rose rapidly during the late 1980s, this was largely a result of owners in the South-east taking advantage of high house prices and moving north to exploit the regional house gap. This is an equilibriating mechanism, although it has now slowed as house prices have stabilized in the South-east and begun to rise sharply in the North. Indeed, there may well be a link between the two phenomena as migrants from the South-east push up prices in the rest of the country.

All this assumes, of course, that labour is relatively immobile. This is only partially true. There is a large group of corporate employees who are moved frequently by their employers and whose relocation costs are paid by the firm. This group is an exception to the general pattern of inter-regional immobility. Most housing market and employment moves take place within the local area with a distance of only a few miles. Longer distance residential moves often lead to long distance daily commuting, at least in the short term, until suitable local employment is found. Inter-regional job migration necessitates change of work and home, and as MacLennan and Jones (1986) has shown, this can lead to paying over the odds for accommodation or to short-term residential moves until the migrants are able to undertake a full local search of the property market.

And some long distance moves are not employment linked at all. In the case of retirement moves, there is the possibility of much greater residential mobility than hitherto. In these cases, the spatial labour–housing market link has been terminated. Conversely, there is another group of highly mobile people; skilled entrants to the labour market who may move hundreds or thousands of miles in search of an appropriate first job. But, once they are plugged into the local housing market, this mobility is generally reduced.

The assumption underpinning much of the preceding discussion is that individuals are in full-time paid employment and limited in their choice of home and work to spatially defined labour and housing market areas. But this is a restrictive assumption which leaves out of the account a significant proportion of the population. Moreover, it also loses sight of the different forms of work available both within and outside of the formal labour market, and the different kinds of household that make up the housing market.

Problematizing the market

Another reason for examining the relationships between jobs and housing is that the contemporary significance of the relation – socially and economically – takes us beyond the confines of the market-place as traditionally conceived. In discussions of the labour market, not only are many accounts restricted to full-time employment, they are often gender-specific. More accurately, they tend to assume a male, full-time wage earner. Similarly, in accounts of the housing market it is often the tenure structure of housing – owned or rented – that holds our attention. In consequence, attempts to conceptualize the links between jobs and housing are conducted through the frame of male wage earners, full-time employment and housing tenure. Such a focus is not altogether surprising, given the significance of this three-way relation. But if we are to take seriously the importance of the links between home and work in people's *daily lives*, then the focus is deficient in a number of ways.

First, as noted above, it loses sight of the different forms of work that people perform in a modern economy. This may range from part-time work to shift work, and from paid work in the informal economy to homeworking. And each kind of work, in different ways, produces a particular kind of connection between jobs and housing. Second, it fails to acknowledge the role of households as mediators of the housing–labour relation, and more importantly, it neglects the sheer variety of household types engaged in formal employment – from the single, one parent household through to the dual-income household. Again, each kind of household experiences the constraints and opportunities of the labour market in different ways. Third, and relatedly, the conventional market focus fails to examine the character of markets – the social relations that constitute them and the structures of provisions that underpin them. As before, we will discuss each point in turn.

One of the most striking characteristics of the postwar economy and especially of the 1980s has been the changing nature of work and its organization. As full-time employment has declined and the notion of a job for life with it, part-time work, especially for women, together with a growing casualization of the workforce have become broad features of economic life. Much of this change is attributable to the decline of manufacturing industry noted earlier and to the growth of service jobs. In many of the expanding service industries, in both the private and public sectors, it is women in part-time, routinized jobs that make up the bulk of the growth. In the more depressed regions of the UK, in those areas where male manufacturing employment has been hardest hit, a woman's wage may be the only income coming into the household. Whereas in the growth regions, especially the South-east of the country, dual-income households have become commonplace to meet the high cost of housing. In both instances, the issue of labour mobility is complicated by taking into account the shifts in the pattern of work and who is actually working.

Nor is it only labour market mobility that is affected by such shifts, so too is the spatial structure of labour and housing market opportunities. As many women work part-time, so that they can organize wage labour around their domestic arrangements such as childcare, their travel-to-work distances are far more circumscribed than those of men. The nature of the difference lies with the sexual division of labour in the household, but the effects constrain where people can work as well as where they can live. Admittedly, for professional women in, for instance, the expanding financial and commercial service industries the distance constraints may be fewer, but the possibility of two 'career structures' within one household raises an entirely different set of constraints around labour mobility and migration. In fact, whichever types of work households engage in, the issue of women's waged and unwaged work alters the very way in which labour and housing markets and their connections are thought through. Our conceptions of work and of the home are likely to be out of step unless they recognize and acknowledge changes in the sexual division of labour within both the labour market and the household.

Indeed, the point can be pushed further to include reference to Pahl's (1984) emphasis upon divisions of labour not only within households but also *between* households. Avoiding the simple equation of work with formal employment, Pahl is keen to distinguish the different kinds of work households engage in, especially those kinds that bind households to the 'informal' or 'hidden' economy. Unravelling these kinds of informal networks also sets the relation between jobs and homes into a different kind of framework (see Watson, Chapter 5).

And much the same can be said for the growth of homeworking in recent years. Difficult to pin down and beset by definitional problems, homeworking is one of those activities that is bound up with structural changes in the organization of work. The shift of work back into the home, especially the kind associated with information technologies, is often taken as an index of more flexible ways of working. The decentralization of work, the increase in subcontracting relationships, the growth of self-employment, the smaller size of firms, and the rise of a 'peripheral' labour force are all part of the speculative trends that situate new technology homeworking as a significant feature of an emergent 'flexible economy'. This is not quite the same as the old 'putting-out' system or the sweated labour that has characterized much of homeworking to date. In fact, the old and the new homeworking divides along class lines, although both kinds entail some degree of job decentralization from factories or offices to people's homes. Whilst neither 'putting out' nor 'telecommuting' nor 'networking' are about to herald the demise of the cities, along with the growth of peripheral kinds of work, they carry a series of implications for how we understand the changing relations between housing and labour markets.

Recognizing the mediating role of households in the changing character of labour market and housing market relations is also important, as much of the foregoing has indicated. For Randolph (Chapter 1), the household is the 'missing link' between the operation of the two formal markets. In his view, households are made up of members with varying labour market capacities, which, in turn, are translated into positions in the housing market. Different kinds of households, therefore, in terms of age, size and composition are likely to possess varied earning capacities and this variation itself leads to diverse housing opportunities. Multi-earner households, for example, are likely to find a very different niche in the housing market from that of single-earner households, even though they may have a similar job and income to those in the multiple household.

Randolph is careful, however, not to reduce his analysis to demography and economics. He also stresses a cultural dimension to household strategies, which includes those of informal work discussed above. Attempting to unravel the kinds of connections between households and markets that he refers to, however, is a complex process. It needs to address, among other things, the kind of approach adopted by Forrest and Murie in their study of owner-occupier households in contrasting housing markets in Bristol in the south-west of England (see Chapter 2). Tracing housing histories, the movement of households through the housing market and relating this dimension to aspects of job history and employment, represents a qualitative way of assessing the dynamics of the two markets and it also has the potential to generate a new set of questions around their connections. For instance, slicing through the labour and housing market in this way may reveal how processes of class formation are integral to the dynamics of the two markets and their interrelations (see Marcuse, Chapter 4).

Finally, there is the unexamined and undertheorized nature of markets in much of the conventional economic literature on housing and labour markets to consider. For example, research conducted along neo-classical economic lines exercises meticulous care over the measurement of such things as the precise impact of rising house prices or the effects of council housing allocation policies on labour mobility. Yet rarely is such meticulous care demonstrated in the conceptualization of the variables used in the statistical models. The concepts of tenure, regional house prices, income and jobs are essentially regarded as unproblematic, as are markets.

Markets, however, are social processes. To operate, they require more than a set of price signals and equilibriating mechanisms. The housing market, for example, involves more than buyers and sellers; it involves developers, builders, banks and building societies as well as exchange agents and private owners. Processes of production, of circulation, and of consumption are all involved in making markets work, and inscribed within each set of processes are economic interests which have to be unravelled in order to grasp what kind of housing is built, where it is built, and for whom it is built. Randolph attempts this kind of exercise in Chapter 1, after having first performed a similar type of analysis on the processes underpinning the labour market. It is for this reason that we have chosen to start the collection with his theorization of the divisions within both labour and housing markets and the links between them. Of all the chapters, his analysis is the most wide ranging, and indeed the most ambitious in terms of setting an agenda for further research on the links between homes and work.

Setting an agenda

The division of this collection into three sections reflects the aim of our agenda; namely identifying, understanding and elaborating the connections between housing and labour markets, or more broadly between homes and work. Indeed, this introduction to the first section has attempted to spell out some of the issues that have to be thought through if we are to embrace the complex relations that hold between the two markets. In that sense, Randolph's chapter is a continuation of that aim in that he too outlines a broad framework of inquiry which summarizes what he considers to be the main aspects of the labour–housing market relationship. In doing so, he draws upon recent developments in labour market theory, in particular the notion of spatially discontinuous labour market segments, in order to rethink the divisions that characterize the housing market and the relations between the two markets. Acknowledging the considerable variation in the ways that housing and labour markets interact in different places and at different times, he attempts to unravel that complexity through an explicit recognition of the role that households perform in connecting the two markets. He also draws our attention to the contingent nature of the connection and the possibility that the two markets operate at different spatial scales. Whereas housing markets tend to 'fix' a population into local housing districts, labour markets appear to operate on a broader scale, at the regional level, shaping the parameters of housing demand. At the very least, he is attempting to sketch out some of the directions that the study of the interconnections may take.

Part Two takes up that challenge. Marcuse and Watson explore, respectively, the class dimension and the gender dimension of the two markets. Different in approach as well as in substance, the two contributions build upon some of the strands of thought set up by Randolph. Equally specific in their foci, the chapters by Forrest and Murie, and by Salt attempt to unpack some of the connections between housing and labour mobility. The two differ, however, in the methods chosen to reveal the connections and the scale of their analysis; one local, the other national. Behind all of the chapters in this section, however, there is an assumption that the connections between housing and labour markets need to be understood more fully, despite the contrasts in the ways that they approach this aim and the different countries from which they draw their evidence.

Part Three takes its cue from the contingent historical and spatial connection between housing and labour markets. Coombes *et al.* provides a statistical analysis of the relationships between house prices and labour market influences across local labour markets in Britain which highlights the connection between buoyant local labour markets and high house prices. This chapter provides a geographical backdrop against which the other chapters examine a variety of relations between labour and housing markets in different parts of Britain. Hamnett focuses on London because of its dominant role in the national spatial division of labour, and examines the links between the changes in its industrial and occupational structure, the structure of the housing market and social polarization. Hudson looks at the North-east – a peripheral region which plays a very different role in the spatial division of labour, focusing on the role of the state in the restructuring of both labour and housing markets and the extent to which state policies have helped produce a divided society. Barlow and Savage also examine the question of social divisions and local policies, but in a very different context, that of Berkshire, one of the most prosperous areas of Britain.

Finally, Goodwin looks at the role of state policy in restructuring labour and housing markets in Docklands, focusing on the extent to which the regeneration of Docklands has been achieved by replacing existing residents and local industry by a new population and new growth sectors. Taken collectively, the chapters in this section highlight the variety of different connections between housing and labour markets in different parts of Britain.

References

Booth, C. (1902), *Life and Labour of the People of London.*

Goddard, J. B. (1983), 'Structural change in the British space economy', in J. B. Goddard and A. G. Champion (eds.), *The Urban and Regional Transformation of Britain* (London, Methuen).

MacLennan, D. and Jones, L. (1986), 'The impact of North Sea oil development on the housing market', in *Land Development Studies*, 3, pp. 113–26.

MacLennan, D. and O'Sullivan, A. (1986), 'The housing market costs of long distance migration'. Discussion Paper 10, Centre for Housing Research, University of Glasgow.

Pahl, R. E. (1984), *Divisions of Labour* (Oxford, Basil Blackwell).

Steadman Jones, G. (1971), *Outcast London* (Oxford, Clarendon Press).

1 *Housing markets, labour markets and discontinuity theory*

BILL RANDOLPH

Introduction

Despite the apparent increase in interest in the relationship between housing and labour markets, relatively little effort has been made to fully conceptualize how the two are related or to develop a framework within which the links between the two may be explored in particular situations. Those studies which have looked at this relationship have tended either to concern themselves with particular areas where the links are more or less explicit, or they have simply juxtaposed empirical analyses of the two processes, usually at the regional level or above, without fully exploring the links between them. An example of the former would be the literature on the 'residualization' of the social rented sector, and of the latter, exposés of the 'north–south' divide in housing costs. However, explanations of how the two processes are related is likely to be found only by studying the links at the local level.

The first objective of this chapter is to stimulate the debate by proposing a conceptualization of the general relationship between housing and labour markets which could form a basic framework with which to analyse their interaction in particular situations. This is not to put forward any grand theory, but rather to attempt to rationalize the complex processes under consideration.

At the same time, the conceptual understanding of how housing markets themselves are structured is also relatively poorly developed. True, we know a great deal about the various divisions in the housing market, both between and within tenures, yet these tend to be empirically, rather than theoretically, defined. Those studies that have invoked theory to define housing market structure have tended to result in rather broad categorizations which lack flexibility or apply only to specific times or places.

The second objective of the chapter is, therefore, to address this gap in our theoretical understanding by assessing the extent to which concepts developed in labour market theory, in particular those of segmentation and discontinuity, can be extended and applied to an understanding of housing market processes. It is suggested that these may provide a more flexible and adaptable framework within which we can comprehend the range of processes – economic, social, political, cultural – that constitute the housing market and the range of outcomes these processes produce. Again, the discussion is speculative; an invitation to debate, rather than a definitive statement or formulation.

The context

First, however, we need to briefly consider the context for the following discussion. Why is an understanding of the relationship between housing and labour markets important and in what ways are these two sets of processes related? The current focus on the problems of reduced labour mobility and skill shortages in southern England, the demand for building land in Green Belt locations, and the increase in homelessness, particularly in London, which have all been seen as a repercussion of the emerging gap in housing market performance between the north and south of Britain, have helped to stimulate a debate among academics, policy-makers and others on how housing and labour markets interact. That there are clear links between the house price boom in London and the South-east and the relative strength of this region's economy, as expressed in terms of both the number and composition of the labour force compared to the rest of the country, cannot be denied. It is, therefore, important to explain these links not only for the purpose of understanding current patterns of urbanization and regional development, but also for current and, more importantly, future policy formulation. Employment creation, population change, household formation and housing provision are key elements in determining planning and investment policies.

However, the awakened interest in the relationship between housing and economic development is merely the most recent recognition, dating back to at least the middle of the last century, of the way in which housing and economic structure are closely interrelated at a variety of spatial scales. From Engels' analysis of Manchester in the 1840s, various studies have attempted to quantify or understand how housing conditions in a wide variety of spatial locations and social situations have been related to their wider economic contexts. Booth's classic study of life and labour in London in the 1890s was essentially a detailed analysis of housing and labour market conditions of the time. More recently, the focus on the inner cities has also exposed the close links between economic decline and falling standards of housing provision in these areas.

Clearly, housing and employment structure – homes and jobs – are closely and reciprocally linked. Where people work and what jobs they do critically affect their housing requirements. Similarly, where people live and how much it costs them, influences both the supply of a suitable workforce, and hence employment location, and the level of wages they will demand. But how are jobs and homes actually related in a functional way? To be more specific, how does the position of an individual in the labour market get translated into his or her position in the housing market?

At the basic level, housing and labour markets are linked through the ability of households to pay for varying quantities and qualities of housing in various locations. This ability to pay is essentially a function of the householder's position in the labour market, which gives him or her the economic wherewithal to afford housing. Household income, therefore, forms the functional link between these two markets. Thus any real understanding of how housing markets work requires an understanding of the way in which the labour market influences levels of household income and hence levels of housing demand.

Over time, however, this link has become more and more indirect through the increasing role of the state. In mediating the outcome of the housing market by redistributive intervention, state action has broken into the operation of a 'pure' market relationship. This has been achieved both in the housing production process, through housing subsidies and the planning system, and by increased intervention in the field of income support provisions and taxation arrangements. Even in those parts of the housing system where market allocation processes dominate, the relationship is made more complex by the operation of a variety of mechanisms that control access to particular sub-markets, for example mortgage eligibility criteria and the local knowledge networks. These also blur the direct links between household income and housing consumption.

However, there are other influences which intervene in the relationship which complicate the picture further. As I argue below, the structure and function of the household itself plays a crucial role in shaping housing demand. Households represent aggregations of individual labour market participants, thus breaking the direct link between individual labour market position and housing consumption potential. Moreover, patterns of intergenerational wealth inheritance and other sources of household capital such as those generated within the housing market itself through capital gains, also affect housing consumption potentials. And wider social and cultural influences, which are focused through the household and are expressed in different patterns of choice and preference, play a vital role in mediating the outcome of the housing market–labour market relationship. These also need to be understood if the link between homes and jobs is to be fully exposed.

Unfortunately, one of the problems in developing an understanding of these links is that the current level of conceptualization of housing market structure is itself limited. Theory in housing studies tends either to consist of applications of marginal economic concepts to housing, or to involve detailed theorization about the structure of housing tenures, state intervention or grandiose concepts of the development of the built environment and the urbanization process. What appears to be lacking is an adequate concept of the way in which the housing market is structured that goes further than simple rationalizations of the various empirical divisions defined in specific locations. Before we can adequately understand the broader relationships between housing and labour markets we need, therefore, to develop a much more flexible conceptualization of housing market structure that can be generally applied to a wide variety of situations. In short, what is needed is a general theory of housing market stratification.

In the following two sections an attempt is made to formulate a conceptualization of housing market structure to fill this theoretical gap by drawing on a discussion of theories developed over the last two decades which account for perceived divisions in the labour market. These have evolved from the initial perception that labour markets were divided into two basic sectors – a 'dual market' reflecting divisions in a range of behavioural attributes and levels of skill and remuneration – through to more complex 'multi-segmented' and 'discontinuous' models in which the processes of labour demand becomes the dominant structuring mechanism. The objective is to identify the key features of labour market structure that produce patterns of labour market stratification. An account of housing market stratification will then be developed by adapting these key features to the context of the housing market.

This discussion also serves an additional purpose, however. Applying concepts derived from labour market theory to an understanding of housing market stratification allows the possibility of developing a unified conceptualization of the housing market–labour market relationship which is based on a common approach to understanding both sides of that relationship. The nature of this relationship is addressed in the final section.

Theories of duality, segmentation and discontinuity in labour markets

Notions of labour market duality emerged from work which attempted to explain the racial conflicts experienced in urban riots which erupted in major US cities in the late 1960s and the decline in inner city employment which underlay them. Early dual labour market theory focused on the supposedly behavioural attitudes shown by different sections of the urban labour force which led to a division between workers in 'primary' higher status, higher income and stable employment and those in 'secondary' lower status, lower income and unstable occupations with few career opportunities. For the ethnic minorities, this situation was reinforced by employer discrimination (Gordon, 1972).

The concept of the 'internal' and 'external' labour markets followed. The former referred to skilled workers in the stable core workforce within a firm. These were retained over time and experienced a structured career progression. The latter comprised workers in unstructured secondary positions who were hired and fired depending on the immediate requirement for labour (Doeringer and Piore, 1971; Bosanquet and Doeringer, 1973).

Notions of a simple duality in labour markets were developed further by more radical segmentation theories. This stressed that labour markets are structured by the demands for different types of labour by employers, not by the attitudes or skill levels of individual labour market participants, and that segmentation of the workforce allows management greater flexibility and control over its workforce. Edwards *et al.* (1975) proposed a three-way model which distinguished between routine 'subordinate' and creative, self-initiating 'independent' occupations within the primary workforce, in addition to the secondary sector. Segmentation theory has subsequently been developed by a number of writers (for example, Friedman, 1977; Piore, 1982; Gordon, Edwards and Reich, 1982) in which concepts of worker resistance, unionization and bargaining on the one hand and divisions between core-periphery and monopoly-competitive sectors on the other have been added to the basic model to produce intra- as well as inter-sector labour market divisions.

More recent attempts to develop models of labour market segmentation have moved away from the more simple two-, three- or four-way typologies. This has led to the notion of 'discontinuous labour markets' where the stress is placed on the conflict between labour and capital as the central feature of labour market structure (Kreckel, 1980; Cooke, 1983). The degree of labour organization is seen to be of major importance in this relationship, with unionized workers being resistant to attempts by management to impose unfavourable working practices and wage settlements. Non-unionized workers are, by default, excluded from the benefits which stem from organized bargaining power. Discontinuity theory, therefore, breaks from the rigidity of dualist or segmentation theories by stressing the importance of bargaining and purposeful action on the part of the workforce in shaping labour market outcomes. It, therefore, introduces a much greater degree of flexibility into the analysis of labour market divisions.

Kreckel (1980) has brought together elements of social stratification theory and labour market theory in order to construct an eight-level hierarchy of private sector labour market situations. These range from illegal immigrants at the bottom end to those with educational, managerial and professional qualifications at the top. Based on Weberian notions of social closure and a 'market situation' definition of class position, Kreckel's model stresses five social mechanisms which affect the bargaining (i.e. market) strength of workers and hence their position in the labour market: demarcation, exclusion, solidarism, inclusion and exposure. Of these, only solidarism is seen as being free from employers' direct influence. The rest are viewed as being ways in which employers can control or manipulate the workforce. Nevertheless, the workforce is able to exert varying degrees of pressure to resist management demands, depending on the balance between the bargaining strengths of capital and labour at any one time.

Kreckel's scheme has an advantage over standard segmented labour market models in so far as it is based on the explicit recognition of the relative, and dynamic, bargaining positions of capital and labour at different levels of the employment hierarchy. Indeed, it is these relative strengths which determine the location of groups of workers within the hierarchy. However, Kreckel does not view the position of the two sides of this market as equal. He distinguishes between two levels of asymmetry in the bargaining power of labour and management. The 'primary asymmetry' of the labour market refers to the dominance of capital over labour in all aspects of the labour process. In the long run, management, as the 'suppliers' of jobs, always has the upper hand. But a secondary layer of asymmetry arises on the labour side through the occupational divisions within the labour force which confer varying access to goods and life chances. Those who can call on privileges such as skill, education or unionization are in a much stronger bargaining position.

> If a specific occupational capacity is *monopolized* by a restricted number of incumbents, labour market competition is thereby

> reduced and their bargaining position, *vis-à-vis* their employers, is improved. This is what Max Weber referred to when he used the notion social closure. (Kreckel, 1980, p.530)

Kreckel's essentially instrumentalist approach can be criticized for neglecting the processes which create labour market divisions in the first place. The typology of labour market positions with which he illustrates his model is also presented as a rigid hierarchy, despite his stress on the variable nature of the conflict that lies behind it. Subsequent developments of discontinuity theory have attempted to push explanation into the relations of production itself, to develop a more fluid approach to the delineation of labour market divisions and, importantly, to incorporate the spatial element in explaining labour market structuration.

Drawing on Kreckel's work, Cooke (1983) has proposed a twelve-way classification of 'spatially discontinuous labour submarkets'. Cooke rejects the strict dual or segmented concepts of the labour market, as well as those which assert manipulative management strategies as the dominant structuring force. Instead, he assigns primacy to the dynamic outcome of conflict over the control and rewards of the labour process by both labour and management. In particular, during periods of heightened conflict, management can utilize spatial discontinuities in the labour market to overcome this conflict in its favour. Focusing on industrial restructuring during the current period, Cooke argues that this process should be viewed largely as a means by which large employers have attempted to reduce labour-related constraints on their actions, reduce wages and thereby increase productivity. Small firms also act as an important element in this strategy, both allowing larger firms to subcontract out in response to worker militancy, and in order to decentralize skilled operations to the small firm sector or to exploit peripheral-type competitive sector conditions that prevail there (see also Friedman, 1977; Berger and Piore, 1981):

> Such spatial resolutions to the capital-labour conflict contribute further to the fragmentation and discontinuity which seem to characterize the contemporary structure of the labour market, which yet retains distinctively hierarchical features. (Cooke, 1983, p.560)

Cooke therefore stresses the dynamic nature of submarket structure rather than its rigidity, as well as the crucial importance of space in creating distinct labour market discontinuities. The labour market is not a simple hierarchy of job segments distributed evenly over space. Particular segments are juxtaposed in a variety of combinations in different locations depending on local employment structure.

Moreover, while retaining a basic primary–secondary division, he argues that the boundary between these two sectors is fluid. Additional dimensions leading to segmentation include the division between the public and private sectors, as well as divisions based on gender and racial/ethnic group. The self-employed also feature as a distinctive category of their own. Cooke, therefore, includes a number of important elements left out by others.

He makes a number of points about the nature of submarket categories. First, despite the hierarchical nature of the schema, submarkets overlap both socially and spatially. Moreover, the presence of a particular segment in an area does not exclude the presence of any other. Second, although Cooke suggests that the secondary sector has a tendency to grow over time, presumably through a deskilling process, there are growth submarkets on both sides of the primary–secondary divide. Finally, Cooke suggests that given the current period of labour market restructuring, we are likely to see the primary–secondary division widen over the foreseeable future.

Storper and Walker (1983) have gone further than Cooke in emphasizing the multisegmented nature of labour market structure and the role of bargaining. Their 'labour exchange model' attempts to convey the interactive nature of the labour market, stressing bargaining as a central process, as well as employer discrimination and employee entrapment:

> The labour market is segmented thanks to both the hierarchy of jobs and the efforts of employers to divide workers in order to win a favourable exchange bargain for themselves. . . Discrimination . . . is widespread because of imperfect information, employer prejudice and the advantages gained by employers from creating divisions in the working class and exploiting those who are weakest. Entrapment of workers in certain labour market segments is a common result. But there is the possibility of manoeuvre on both sides of the bargain rather than complete capitalist domination and manipulation. (Storper and Walker, 1983, p.16)

In addition, there are no a priori definitions of segments. Storper and Walker's model, therefore, stresses the contingent nature of historically and social produced labour market structures, and the non-predictable and often contradictory nature of the job creation process. The definition of segments can only be determined in the course of specific empirical analysis.

We have moved a considerable way from the early notions of dual labour markets. The notion of a discontinuous labour market leads to a much more complex conceptualization which emphasizes the role of relative bargaining strengths between participants and throws the stress on the 'market' of labour markets i.e., the changing relationship between the supply of and demand for labour in a variety of forms. Drawing on the discussion above, we can summarize the main features of spatially discontinuous labour markets as follows:

1 The asymmetry of the capital–labour relationship. At the end of the day, it is the employer who hires and fires. The number and type of jobs available are, in the first instance, determined by the requirements of employers. Similarly, if particular jobs become superfluous, for example because they are replaced by cheaper labour processes, falling productivity, or a firm's market share falls, then employees are laid off. In other words, labour market structure is determined by those who 'demand' labour, not from those who 'supply' it. Employees can only 'choose' from those job opportunities that employers choose to offer.
2 The importance of bargaining and conflict to establish the relative strengths of labour and capital. It is in this context that the notion of inclusion and exclusion from 'core' labour market positions is important in establishing the relative strengths of each side. 'Internal' workers and those who can draw on solidarism to strengthen their position enjoy greater bargaining power than 'external' workers excluded from the core workforce and subject to much greater exposure. These variations in bargaining power impart asymmetry into the relationship between segments within the labour force itself.
3 Discrimination and entrapment are important features which help to explain submarket structure, reinforced by the imbalance between employers and employees in the degree of information on labour market conditions and power to exploit market imperfections. Gender and race are obvious elements here, reflecting employer/employee perceptions and attitudes formed by wider social and cultural forces.
4 Labour markets are spatially constituted. Individual segments have very different spatial characteristics, from those that draw on a labour force with internationally exchangeable skills (for example, corporate executives, academics, migrant workers) to those that recruit from very local labour pools (for example, shop workers, domestic workers). Moreover, the uneven opportunities for profitability that different locations offer play a central role in inter-firm competition and hence on the relationship

between firms and their workforce. In effect, space provides management with a tool to outmanoeuvre labour during the bargaining process. However, there is a wide variation in the extent to which employers are able to use space in this way. Transnational corporations are able to exploit labour market differentials at a global scale, while small firms serving local markets have little choice but to recruit from the available local labour supply. There are, then, a considerable variety of discrete labour market segments, some being tightly drawn within a particular locality, while others have a much wider spatial spread operating at the regional, national or even global scales.

5 The model stresses dynamism in the relationship between sub-markets as some expand and others contract over time and space – the notion of leading and lagging sectors, within which differential rates of profitability and labour processes affect the ability of workers to resist conditions that management wish to impose.

6 While the model includes a hierarchical alignment of segments, they are seen to co-exist both socially and spatially in a series of overlapping layers. This reflects the considerable flexibility of labour market structure and does not preclude workers moving between jobs in different segments which demand similar or transferable levels of skill.

7 Although the model recognizes a basic primary–secondary division, there is nevertheless considerable fluidity in the structure of the boundary between them and indeed in the boundaries between segments on each side. The hierarchy of segments is, therefore, far from rigid and is characterized by considerable fragmentation.

8 The model stresses the social, historical and spatial contingency of labour markets and the unpredictable and often contradictory nature of the job creation process. In empirical terms, this implies that there are no a priori categories. The model's value lies in its conceptualization of the processes involved rather than its descriptive potential:

> The aim is to try to fix conceptually the types of job that may be anticipated in various kinds of location, and subsequently, to give brief indications of tendencies towards significant change in such categorizations. (Cooke, 1983, p.556)

It is not my intention to assess the value of spatial discontinuity theory in explaining labour market structure in Britain. However, it is worth noting that an emphasis on space as a crucial element in the structuring of labour market processes is a central theme in recent work on the spatial division of labour. This is based on an understanding of labour market restructuring as being essentially a process of *spatial* restructuring within the economy (Anderson, Duncan and Hudson, 1983; Massey and Meegan, 1982; Storper and Walker, 1983; Massey, 1984). The emerging polarization in economic performance of non-metropolitan South-east England and the North and North-west has been just one, albeit the most recently prominent, manifestation of this spatial restructuring process. In addition to the concentration of new investment in new growth regions, spatial differential in economic growth and decline has also been expressed within the economies of the main metropolitan areas themselves. The polarization of the workforce noted above has been most pronounced in the major cities, particularly those where growth is in expanding labour market segments, particularly in services and office-based employment, and the decline in traditional manufacturing employment has been rapid (GLC, 1985; Hamnett and Randolph, 1986a). In those cities which function as key centres within the global economy, the emergence of specialized labour market segments, especially in the financial and securities sectors, have led to an even greater degree of polarization and new patterns of segmentation (Simmie, 1983; Sassen-Koob, 1985).

Dualization and segmentation in the housing market literature

To what extent can the eight elements discussed above be of use in understanding housing market structure? Before assessing this potential, it is worth briefly reviewing the extent to which concepts derived from or comparable with those applied to labour market theory have already entered the housing market literature. Notions of dualization and segmentation have been around in housing market literature for some time, although generally they have been adopted in a rather vague manner and have not necessarily been framed explicitly in these terms. The concept of dualization can be seen in those studies that have considered the social and political consequences of the evolution of tenure structure over the past quarter of a century. The basic proposition, which has been widely accepted, is that the private rented sector has been replaced by a dualized housing system in which owner occupation and renting from a local authority have become the dominant tenure alternatives. This division has been associated with the diverging social characteristics of those housed by the two tenures in recent years (Hamnett, 1984; Hamnett and Randolph 1988a).

Tenure-based social polarization has been explained as part of a much broader set of social economic and political process relating to changing patterns of control and distribution of resources and power within society. During the current period, this has resulted in an economically and politically 'marginalized' welfare dependent population, heavily reliant on direct state subsidies and increasingly concentrated in a 'residual' state housing sector, in contrast to an economically active population, supported by indirect state subsidies (such as tax reliefs) and other private subsidies (low cost mortgages, etc.), and increasingly associated with home ownership. Thus a 'primary' owner-occupied sector contrasts to a 'secondary' state-rented sector (English, 1982; Ball, 1982; Malpass, 1983; Murie, 1983; Forrest and Murie, 1986).

The comparison with dual labour market theories in this literature is quite evident. Indeed, the dualization of housing market opportunities has been linked to the trends in labour market structure which have produced a greater inequality between those with and those without economic and political power (CDP 1976; Friedman, 1977; Byrne and Parsons, 1983; Harrison, 1983). However, the extent to which these tendencies are the result of changes to the housing system itself, such as shifts in subsidies and changes in the cost and availability of housing finance, as opposed to changes in the economic and social composition of the population housed (i.e., supply rather than demand factors) are often difficult to untangle.

There are considerable doubts, however, as to whether housing market processes as a whole can really be characterized by such a straightforward dichotomy. Intra-tenure divisions are in many cases just as significant as those between tenures (Forrest, 1987; Forrest and Murie, 1987). While the polarization thesis can be substantiated in areas such as inner London, where gentrified private housing and residual council estates exist cheek by jowl in many areas (Randolph, 1987), elsewhere housing market processes are much less extreme in their expression, or if they are, the extremes do not tend to be juxtaposed in such close proximity. Although the polarization thesis refers to a tendency rather than a rigidly observable division, it nevertheless has clear limitations in helping to explain housing market structure as a whole and the complex relationship between housing and labour markets which vary over time and space.

Attempts to develop explanations of this relationship using more complex conceptualizations are relatively limited. Perhaps the most well known is the concept of housing classes developed by Rex and Moore (1967). Their six-way classification of housing submarkets in Birmingham during the late 1960s is essentially a description of a particular pattern of housing market segmentation. While the original theoretical basis of the concept has been subject to extensive criticism and revision (Haddon, 1970; Rex and Tomlinson, 1978; Pratt, 1982; Saunders, 1982), housing class theory does attempt to relate both social structure and housing market structure together in a way that explicitly recognizes that the housing market is multisegmented rather than shaped by a simple duality.

Significantly, the central focus of Rex and Moore's study was the explanation of the particular position of ethnic minorities in the housing market. It is no coincidence that one of the areas of housing studies in which the notion of housing market segmentation is well established (albeit not explicitly expressed in these terms) has been in the race and housing literature (Parker and Dugmore, 1976; Flett, 1977; Ward, 1982; Henderson and Karn, 1984; Karn, 1985). Here, intra-, rather than inter-tenure differences are particularly important. For example, the contrast of low income Asian home owners with the remainder of the owner-occupied market highlights the very considerable differences that occur within individual housing tenures. Similarly, studies considering the position of racial minorities in council housing were among the first to confirm that council housing is itself subject to segmentation. More explicitly, Cross (1983) has utilized notions of labour market segmentation in developing a wider explanation of the residential entrapment and segregation of ethnic groups.

The value of adopting segmentation theory by those trying to understand ethnic divisions in the housing market is not surprising, given that race and ethnicity cut across the more usual concerns about class and tenure. To a large extent, the recognition that tenures are far from homogeneous and that a more subtle understanding of the processes involved is required, parallels the early development of dual and segmented labour market theories, originally aimed at explaining the position of black and other minority groups in the American urban labour market.

Nevertheless, the development of segmentation theory in housing studies has been slow to evolve, despite the fact that the main two tenures are highly differentiated in terms of submarkets and can be defined in terms of their distinctive social characteristics (CDP, 1976; Taylor, 1979; Forrest and Murie, 1983; Twine and Williams, 1983; Clapham and Kintrea, 1986). In general, these divisions have been defined by empirical, rather than conceptual, analysis, although they have usually been explained as a result of differential housing allocation processes. Where the concept of segmentation *per se* has been explicitly invoked, it has been used as an empirical tool (Munro, 1986).

More recently, Forrest (1987) has put forward a three-way division of the housing market that cuts across tenure categories and which more or less explicitly draws on the basic labour market segmentation model. In Forrest's model there are, first, those who form part of the 'internal' core workforce with higher skills and income, often operating in a national or international housing and labour market. Their housing position is heavily aided by occupational subsidies, a high rate tax relief and substantial capital gains. A second category comprises the bulk of the economically active population who work outside the core sector and whose access to an essentially local housing market is dependent on normal channels of housing credit and more modest levels of interest tax relief, and for whom opportunities for capital accumulation are relatively limited. The third group consists of elderly home owners and tenants in both the public and private sectors. Forrest suggests that public subsidies for these households are relatively high but occupational subsidies are nil, and that bureaucratic allocation processes are important and housing opportunities are severely spatially restricted.

Forrest's model explicitly links housing and labour markets together and strongly suggests that an understanding of housing market structure cannot be based on tenure divisions alone: '. . . if we are to escape from overemphasising housing tenure divisions . . . we need to develop a more unified and processual framework in which to situate debates about social divisions in housing' (Forrest, 1987, p. 1,616). Forrest also stresses the role of bargaining for households in different sectors, and introduces the notion of entrapment for those excluded from access to both good jobs and good housing. For example, the first group can exert considerable bargaining strength – and hence choice – in the housing market due to their economic power and can also bargain for the occupational subsidies denied to less powerful sections of the workforce. In political terms also, this group benefits from substantial state subsidies in the form of tax reliefs for housing consumption – part of the 'middle-class welfare state'. In contrast, the political and economic marginalization of the third group is reflected in its poor bargaining position in the housing market and the stigmatized nature of the housing it occupies and the subsidies it receives.

Even this brief and selective review of relevant housing literature suggests that concepts of duality and segmentation borrowed from the labour market literature are clearly valuable in developing a conceptual understanding of housing market processes. Of course, there are dangers in adopting concepts developed to explain one type of social process to an understanding of another. For example, it is not possible to simply adopt the divisions suggested by labour market models and impose them on housing market divisions. Housing markets possess distinctive structures that reflect the autonomous processes underlying the nature of housing production and supply, for example, tenure structure, dwelling type, and so on. These processes largely determine the form in which housing is supplied and have their own impact on housing market structure, quite distinct from those processes operating in the labour market. Any theorization of housing market structure, therefore, needs to accommodate the distinctive roles of both housing consumption and production processes.

Towards a theory of spatially discontinuous housing markets

What, then, can spatially discontinuous labour market theory add to our understanding of housing market processes? At a very basic level it provides us with a more sensitive tool with which to understand that part of housing demand that derives from the position of housing consumers in the labour market. The greater flexibility of the concept, compared to more rigid dual and segmented labour market approaches, provides much greater scope for accurately understanding the way in which housing demand varies both between and within particular localities. By building such concepts into the demand side of the model, we may appreciate further the impact that local labour market structure has on the housing consumption propensities of households, particularly during periods of rapid restructuring. For example, the local impact of the recent recession on the ability of home owners to maintain mortgage repayments appears to have been crucially affected by the presence of alternative growth sectors, particularly service industries employing part-time married women, which may offset the loss of 'traditional' male manufacturing employment. Similarly, the reaction of certain sectors of the Central London residential property market in the wake of the October 1987 stock market crash can only be explained by the presence of a sizeable sector of housing consumers in Central London who work in the City financial markets and who in turn form part of a wider global financial labour market. An understanding of labour markets using simplistic dual or segmented labour market concepts would have been an inappropriate tool to explore both these situations.

But the value of discontinuity theory lies not only with its ability to inform our understanding of the demand side of the housing market equation. It also provides us with a set of conceptual tools which can further our understanding of housing market processes *per se*. Drawing upon the summary of the main features of spatially discontinuous labour market theory outlined earlier, the following elements may usefully be incorporated into a theory of spatially discontinuous housing markets.

1 Asymmetry of the relationship between producers and consumers of housing. In a market situation the relationship between producers and consumers is a reciprocal one (neither can function without the other) and both sides of the 'market' influence the actual outcome. But it should be emphasized that housing, like employment, is structured in specific ways by the requirements of or constraints upon those agents and institutions concerned with the supply of housing, rather than the product of consumer choice or demand. This is not to deny that the demand for housing has an influence on housing supply and that the balance of this influence changes over time depending on wider market conditions. But it does mean that the key decisions about what form of housing is produced where and for whom lie with housing producers – builders, developers, landlords, financial intermediaries, housing departments, policy-makers, etc. – and not housing consumers. In other words, an understanding of housing market structure should start with housing production, not housing consumption. Such an approach stands in contrast to Rex and Moore's (1967) housing class model which focused essentially on housing consumption rather than housing production. It also provides a strong antidote to 'consumer-led' concepts of market relationships.

A 'supply-side' approach to understanding the origins of housing market segmentation not only refers to the physical aspects of housing provision – the numbers and location of housing, its type, size and the standard of provision, etc. – but also to its institutional form; in particular, its tenure characteristics. Housing tenures form key lines of division in the housing market because they represent the structures that ensure that housing can be produced and sold at a price that households can afford while at the same time guaranteeing a profit for the builder and safeguarding a return on the finance advanced to fund the consumption process (Ball, 1982). Tenure is therefore an essential element of the housing production process.

Moreover, the form of tenure through which housing is consumed has a very important impact on the relative bargaining strength in the housing market of these consumers. For example, under private landlordism the use and investment values of housing were split between the tenant on the one hand and the landlord and private mortgagee on the other. But under the owner-occupier system the home owner takes on the role of both landlord and tenant.

The significance of this is that owner occupiers stand in the same position to the investment value of their properties as do landlords, and gain as the investment value of the property appreciates. This gain is denied to the tenant. The differential ability of tenants and home owners to benefit from what has become the major domestic investment asset, and hence enhance their bargaining strength in the housing market *vis-à-vis* tenants, is today a major line of division within the housing market. Indeed, it has major ramifications well beyond the housing market itself, influencing social and political attitudes and labour market potentials. The closest labour market equivalent is the contrast between employees and those in self-employment where the 'ownership' and use of labour refer to the same person. However, the latter form only a small minority segment.

From both the production and consumption sides, therefore, tenure divisions *are* highly significant in understanding housing markets and introduce a major element of discontinuity into the segmentation process.

This discussion highlights the point that housing markets involve very distinctive processes which impart distinctive lines of demarcation in the segmentation of housing market structure. A central task, therefore, is to isolate those elements of housing market segmentation which are a result of the housing production process (including the land dealing system, housing construction, the housing finance institutions and state subsidies which flow to house production) from those that result from the way housing is consumed. The latter refer primarily to the differential market power of households, derived from their labour market position and various housing consumption subsidies and income support measures.

2 Despite this primary asymmetry, the role of bargaining between producers and consumers is essential to an understanding of housing market situations. Households are able to exert variable pressure on those supplying housing owing to their varied financial capabilities derived from the position of household members in the labour market, their ability to benefit from state and other subsidies for housing consumption, and their access to capitalizable assets, particularly through previous ownership or inheritance of domestic property, which can be used in housing transactions. On the other hand, the bargaining strength of those who supply housing services also varies, depending on the external factors which affect their operations. The ability of builders or developers to extract higher prices depends to a great extent on the general reaction amongst consumers to prevailing market conditions or the extent

to which scarcity can be exploited. There are buyer's as well as seller's markets.

Notions of inclusion, exclusion, demarcation, exposure and solidarism in relation to differential access to housing services, also have a direct relevance to understanding how 'bargaining', and hence the role of purposeful action, works in housing market processes. Such concepts are well developed in the urban sociology literature. Bargaining here refers not only to the relationship between housing producers and consumers, but also to the relationship *between* households, where differential market ability to compete for housing represents a major line of demarcation. This inter-household competition represents a secondary asymmetry in the housing market; one which reflects differential demand propensities rather than differences in structures of provision.

For example, the notion of inclusion could be applied to households who are able to access preferred or privileged housing market segments due to superior labour market strength. Exclusion would apply to households who are unable to gain access to any particular market segment: for example, exclusion from home ownership due to low pay, unemployment, welfare dependency, age and so on. On the other hand, the single homeless are often excluded from social housing owing to their failure to qualify as priority needs cases. In this case, exclusion stems from the structure of housing provision, not from differential demand propensities.

Demarcation in housing markets could refer to the ability of housing professionals, such as estate agents and mortgage managers for example, to manipulate households into selected segments. Solidarism could be applied to situations where groups of households combine to resist threats to their position, either from housing producers or other competing households. These latter features are largely a product of attitudes and perceptions formed within the wider social and cultural environment and outside the housing market itself. In this way, wider social behaviour influences the bargaining power of households in different market segments.

3 Following on from this last point, the ability of certain households to compete in the housing market is clearly affected by discrimination, most obviously on grounds of race or ethnicity. Discrimination along gender divisions is also significant here and in certain locations other cultural and religious divisions may be important. Again, these lines of demarcation represent the outcome of social processes which are essentially formed outside the housing market *per se*. Entrapment also has a clear relevance to understanding housing market processes, especially for those segments where choice is limited and bargaining strength is weak. The notion of

housing market entrapment has played a significant role in explaining the concentration of disadvantaged groups in inner city areas, particularly for those in residualized council housing.

4 Space plays a crucial role in housing market processes, quite simply because housing involves the consumption of land. The housing market is a spatially constituted market *par excellence*. It is therefore impossible to understand housing market processes without recognizing space as a fundamental component of these processes. To begin with, each segment in the housing market has its own spatial characteristics. For example, the part of the housing market that caters for mobile executives may operate at the national level, reflecting nationally (or, indeed, internationally) determined income scales and relocation payments. On the other hand, the local authority housing market, by definition, is largely restricted to local authority areas. Such segments are therefore relatively limited in spatial terms and largely local in character.

Second, while housing market segments tend to be differentiated broadly in space due to the way in which the housing system has developed over time (for example, until recently, owner occupation was seen as a feature of the post-1930s suburbs while private renting was concentrated in the nineteenth-century inner cities), they nevertheless overlap geographically. Housing market segments are not therefore spatially exclusive. It is this feature which imparts the high level of *spatial* discontinuity into housing market structure. A particular locality could be characterized by a variety of interpenetrating housing market segments.

The housing market in any one locality is thus characterized by a hierarchy of overlapping socially *and* spatially defined segments. It is not a unitary hierarchy, however. Rather there are a number of tenure-based divisions that intersect with local social divisions to produce a range of market segments. Each segment will have its own distinctive economic, social and locational characteristics, defined both in terms of the structure of provision on which it is based and the social characteristics of the population it accommodates.

5 We need to understand the housing market as a dynamic process, one that is constantly changing in response to changes in the structures of provision and the characteristics of demand. The character of housing market segmentation not only varies over space, but also varies over time. Different segments emerge and decline depending on both changes in the structure of housing provision (changes in taxation, investment conditions of legislation, for example) and changes in the structure of demand. The expansion of a long

leasehold owner-occupied flat sector in British cities since the 1950s is an example. Moreover, the slow evolution of the housing system means that in any one period remnants of earlier tenure forms remain long after the rationale for them has passed. The long demise of the traditional unfurnished rented sector is perhaps an obvious example. On the other hand, new segments may exist in embryonic form for a considerable time before they become significant. In Britain, housing associations are a clear example of this. These 'emergent' and 'relic' segments form an important element in the evolution of housing market discontinuities.

Moreover, the fact that housing is a spatial commodity means that we can view the various cycles of investment and disinvestment moving across the surface of the housing market in a continual process of revaluation and devaluation of specific segments. At any one time investment in certain markets is expanding while in others it is receding. Gentrification and residual council estates are contemporary tenure specific examples of both sides of the investment-disinvestment divide in Britain. At the broader level, the emerging north–south divide in patterns of investment and valuation in the owner-occupier market is another. Clearly, the housing market has leading and lagging edges just like the labour market. The segmented structure of the housing market is therefore highly dynamic, both over time *and* space.

6 While recognizing that each segment possesses a degree of autonomy in terms of its internal function and consequently the market it serves, the boundaries between segments are not rigid. Households usually have the ability to move between a number of segments at any one point in time, depending on their 'bargaining' strength and market position. The autonomy of each segment is therefore relative. Major lines of discontinuity may exist between the main tenures, for example, or between different social segments which may remain relatively impermeable. We have already noted that at the present time there is a tendency towards a central division that mirrors the primary–secondary division of the labour market and is drawn roughly between home ownership and renting. But within each side of this main division, individual segments may be highly fragmented into a range of interpenetrating submarkets. Nevertheless, even the primary–secondary division between home ownership and renting is a relatively fluid one.

7 Lastly, we need to stress the social, historical and spatial contingency of housing market processes. Although the major discontinuities in housing market structure can be understood at the aggregate level, the detailed segmentation of local housing markets

are only definable through specific empirical analysis. The main implication of this is that the outcome of apparently similar sets of processes will vary between localities and over time because of the variety of historical and social contexts within which they are embedded. Moreover, patterns of housing market segmentation are not fixed forever, but constantly evolve.

This conceptualization of housing market structure, therefore, does not lead to the a priori specification of housing market segments *per se*. What it offers are a number of key elements that should be considered in an explanation of housing market segmentation in any particular locality. Using these elements as a framework, it would be possible to tease out those processes leading to distinctive patterns of segmentation operating on both the supply and demand sides of the housing market equation.

Towards a framework for the analysis of housing and labour market interrelationships

The discussion of discontinuity theory in relation to housing and labour market structure leads on to the second objective of this chapter; namely to develop a framework to help understand the way in which housing and labour markets are interrelated. As the basis of this relationship and some of its current manifestations have been discussed at some length elsewhere (Hamnett and Randolph, 1986b), I shall limit the discussion here to establishing how the formal links between these two markets may be portrayed. In doing so, the role of the household in mediating this interrelationship will be considered.

Complexity and variation in the housing–labour market relationship

Quite clearly, it is impossible to simply 'read off' the housing market position of an individual from his or her labour market position, or vice versa. There is simply a too great a degree of complexity in this relationship. Where does this complexity come from? In order to answer this, we can begin by noting four characteristics of the housing–labour market relationship which tend to operate at the aggregate level.

1 Labour markets and housing markets exist in a reciprocal relationship due to the fact that the former constitutes a major element of the demand side of the latter and the latter represents the repository of the supply of individuals for the former.

2 Despite this linkage, housing and labour markets are, to a large extent, relatively autonomously constituted processes. The mechanisms that generate and structure the two markets involve essentially distinct sets of processes. There is, therefore, no necessary correspondence between the two markets. Instead, the relationship between them must be viewed as being spatially, socially and historically contingent. This imparts a high degree of locational and historical variety in the outcome of the relationship.
3 The two markets are not only relatively autonomous, but also relatively non-synchronous in the sense that changes in one may bear little necessary relationship to changes in the other, and indeed may be contradictory. For example, changes in housing policy leading to greater home ownership may coincide with a period of high unemployment. Moreover, as housing is the more stable of the two, it exerts a frictional drag on labour market change. Pools of 'redundant' labour are liable to get left behind when employment opportunities collapse or move elsewhere.
4 The two markets operate at rather different levels of spatial resolution. To put it crudely, labour markets tend to operate at the regional level, structuring the broad pattern of housing demand, while the outcome of the housing market is much more locationally specific, 'fixing' the population into socially definable residential districts.

The last three characteristics in particular, result in considerable variety in the way in which housing and labour markets interact in different localities and time periods. But there is another layer of complexity in the relationship which does not simply stem from the characteristics of the two markets. While there is some predictability between the position in each of these markets for the principal or sole wage earner in a household, for other household members or for those households comprised of individuals outside the formal labour market, the situation is much more complex.

This last point gives us a clue, for it can be argued that much of the complexity in the labour–housing market relationship results from the composition and nature of those households through which the relationship is articulated. Households are themselves an essential element of the labour–housing market relationship due to the fact that households contain both housing consumers and labour market participants. The important point here, of course, is that it is *households* who consume housing, but *individuals* who participate in the labour market. In other words, understanding the labour–housing market relationship requires explicit recognition that this relationship is articulated through individual housing and labour market participants and the households in which they live.

The household – a missing link?

The role of the household in the housing–labour market relationship can be simply stated. As combinations of individual labour market members, households represent a major disjuncture in the relationship between labour and housing markets by mediating the way in which the market capacity of individual labour market participants, derived from their employment status, is translated into a position in the housing market, which is derived from the consumption potential of the households in which they live. The aggregation of individual labour market participants into households creates an almost infinite variety of household types whose housing needs and consumption potentials bear no direct relationship to the structure of the labour market in which each individual is located. Hence the impossibility of simply 'reading off' the housing market position of an individual from his or her labour market position.

What does this mean in practice? At a very basic level, the size and composition of a household directly affects its housing consumption in terms of the physical aspects of housing demand. As a result, individuals in similar labour market positions will have very different housing demand potentials depending on the demographic characteristics of their households.

The economic characteristics of households will also have a major impact on housing consumption potentials. Individuals in multi-wage households are likely to be in very different housing market positions to those with identical occupations and incomes, but in single earner households only. Similarly, the position of an individual in his or her household will have important repercussions on the housing market position they are likely to be in. Household composition also affects the way in which an individual's housing market position interacts with their labour market potential. Individuals in a similar housing situation, but with different household characteristics, may have very different propensities to change jobs or move to new jobs. Those with family responsibilities, such as children at school, may be much more constrained in their labour market choices.

Thus the economic and demographic composition of households and the economic relationships between household members actively mediates the interrelationship between housing and labour markets, reducing the direct correspondence between them in terms of the labour market capacity and housing consumption potential of any one individual.

However, it can be argued that the impact of household structure on housing and labour market relationships is much more profound than simply one of economics or demography, for the household represents a separate sphere of social activities and relationships that are relatively autonomous of either housing or labour market position. These reflect a wide range of social and cultural relationships that operate at a broader societal level. For example, household work strategies, particularly where they include informal economic activity, as well as the provision of informal housing services such as house sharing and renting rooms to lodgers, may have a significant impact on the attitudes and reactions of individual household members to formal housing and labour market opportunities. The home can function as a source of actual income where services are provided for others or as a source of imputed income as services which otherwise would have to be bought in are provided on a self-provision basis (Gershuny, 1978, 1983; Pahl 1980, 1984, 1985; Redclift and Mingione, 1984). Pahl, in particular, has shown that the home can function as a location where a wide variety of consumer-based activities are carried out which represent, in effect, a substitute for work performed in the formal labour market. Homeworking occupies something of a midway position between the purely informal and purely formal labour markets, which depends on the home as a working location where the overheads of the formal labour market can be avoided by employers (Hakim, 1982; Donnison, 1985; Evans and Cooke, 1985). Thus there are a wide range of complex activities that take place within the context of the household and which function as alternatives or supplements to any livelihood gained from the formal labour market. These may critically influence attitudes and perceptions that individual household members may hold towards both the housing market and labour market, and their capacity to compete in either.

Our understanding of the role and importance of the home as a source of informal housing services is much less developed, however. Nevertheless, anecdotal evidence suggests that it may well be of considerable significance, especially for certain households at particular periods of their housing cycle. Households in the early years of home ownership may enter into informal renting arrangements to provide a source of extra income to offset mortgage payments. This certainly seems to have been the case for immigrants (Burney, 1967, Rex and Moore, 1967). More recently, survey evidence from a recent study in inner London suggested that up to a quarter of owner occupiers in a gentrified neighbourhood gained some form of income from renting out a part of their home, particularly on a house or flat share basis (Randolph, 1987; see also Allen and McDowell, 1989).

As the site of these informal economic activities, therefore, the home takes on an additional meaning over and above its use as shelter or the location of domestic life. In some cases, they will be sufficiently important that the relationship between an individual's position in the formal labour and housing markets will be significantly affected. The extent to which these informal activities modify and influence this relationship and, in particular, the use of the home as a source of informal housing opportunities, seems an area worthy of greater research.

But as well as being important in terms of its informal economic potential, the household is also significant in terms of the social and cultural practices which structure household behaviour and action, particularly in terms of the symbolic use of the home over and above any considerations of shelter and investment (Saunders, 1982; Saunders and Williams, 1984; Williams, 1986). There is a growing body of work which has explored the role of housing as a status object and an arena for personal expression and autonomy as well as the site for a major part of the socialization process (Agnew, 1981). The status values placed on housing by different households contributes in no small way to decisions taken on where to live and in which sort of property. Such attitudes cut across purely economic considerations. Status and fashion, therefore, can be very potent forces in shaping housing market processes.

Indeed, a whole range of new behavioural patterns based on novel lifestyle preferences, even if nothing more than the result of home-based consumption stimulated by modern mass-marketing techniques, are playing an increasing role in influencing housing choices, especially for the expanding middle to higher income labour force. Thus an explanation of gentrification needs to account for the emergence of new, media-inspired lifestyle preferences which have stimulated this process – 'period' restoration, designer furnishings, 'amenity-rich' inner city environments, etc. – as well as the emergence of a high income urban labour force and housing market factors (tax and subsidy changes, etc.) that have encouraged the consumption potential of these workers to be invested in devalued inner city housing (Ley, 1981; Smith and Williams, 1986).

The household is also a sphere of social activity where gender relations assume considerable importance, not least in the relationship between the formal and domestic economies through the dual role of female work careers. The movement of women into and out of the labour force for child rearing or for other domestic needs (caring for elderly relatives, for example) may have a significant impact on the housing consumption potential of their household. The importance of child rearing as a distinctive stage in the family life cycle and its impact on housing careers has been well recognized in urban residential theory for many years, albeit merely in relation to its effect on housing choice (Shevsky and Bell, 1955; Foote, *et al.*, 1960; Johnston, 1971). In recent years, the emergence of a feminist perspective on the domestic relations of the family and the home have opened the way to a more significant appraisal of this aspect of domestic life (Pratt, 1981; MacKensie and Rose, 1983; Bowlby, Foord and McDowell, 1985; Watson, 1985).

And as the major arena of domestic social relations and potential domestic conflict, the household exerts a major independent influence on both the demand for employment and housing as a result of changes in the rate and form of household fission and formation. Here, changing social attitudes to marriage and family life as well as the growth of single person households, both as a result of increasing numbers of single elderly on the one hand and an increasing tendency for the young to lead independent lives before marriage on the other, are important. Such changes in social behaviour can, of course, be related to changes in the labour market, for example through the greater social mobility brought about by the growth of higher paid professional and managerial employment, the increasing role of women in the labour force and the tendency over the last quarter of a century to move away from home for both education and employment opportunities. Although economic factors may well underpin such changes, it is likely that an additional impetus comes from changes in social attitudes to the home and domestic arrangements over and above these purely economic factors.

What this discussion suggests, then, is that there is a wide range of social relations and practices, which find their expression within the household, that can have a major impact on modifying the underlying economic relationship between individual household members and their positions in the formal labour market on the one hand and the formal housing market on the other. The sphere of reproduction, in the form of the household, could be said, therefore, to represent something of a 'missing link' in the housing–labour market chain. Before we can adequately understand the outcome of the housing market–labour market relationship, we need to explicitly recognize and explore the impact on that relationship at the level of the individual of those social and quasi-economic processes which are mediated through the household and the domestic economy.

The relationship between housing and labour markets: a conceptual model

We are now in a position to make an attempt at specifying the way in which housing and labour markets are interrelated. This is set out in Figure 1.1, which draws together the foregoing discussion in a simple schematic framework that summarizes the main components of the labour–housing market relationship. The framework outlines the major elements that need to be considered in explaining the relationship between housing and labour markets in any particular location. It also stresses the links between the spheres of production, consumption and reproduction through which the labour–housing market relationship is articulated, and highlights those areas in the relationship where discontinuities are introduced. These include the importance of space in structuring both labour and housing market segmentation, and the disjuncture in the relationship caused by the household.

On the labour market side, jobs, created by the labour demands of firms and other organizations and agencies in both the public and private spheres, broadly constituting 'the labour process' in its most general sense, are matched to the supply of labour in the form of individual members of the labour force. The interrelationship between the supply and demand for labour is mediated by the operation of the allocative properties of the labour market which 'sifts and sorts' the labour force into particular jobs. The social dynamic in this process is exerted through the relative bargaining positions of employers and the labour force, which will reflect the social, political and cultural milieux in which bargaining takes place.

The result is a series of spatially discontinuous labour market segments which will vary between localities and over time. These will be defined by the interaction of the particular characteristics of labour demand in the area and the nature of the available workforce. The labour market in this case is taken to include *all* potential participants in the labour process, including those who may not play an active role (for example, women with young children, the retired, etc.). The latter may well, however, play a central role in those informal economic activities, especially within the domestic economy, which can and do modify the basic market relationships involved. They also represent a pool of potential labour which may be drawn into formal economic activity from time to time depending on changes in the structure of production or the nature of the existing labour supply.

The position of individuals in the hierarchy of segments is the major determinant of their ability to consume housing services, as this will define their basic economic potential. Any empirical analysis of labour–housing market relations, therefore, needs to explore the processes operating to structure the local labour market in order to understand the underlying economic determinants of local housing demand.

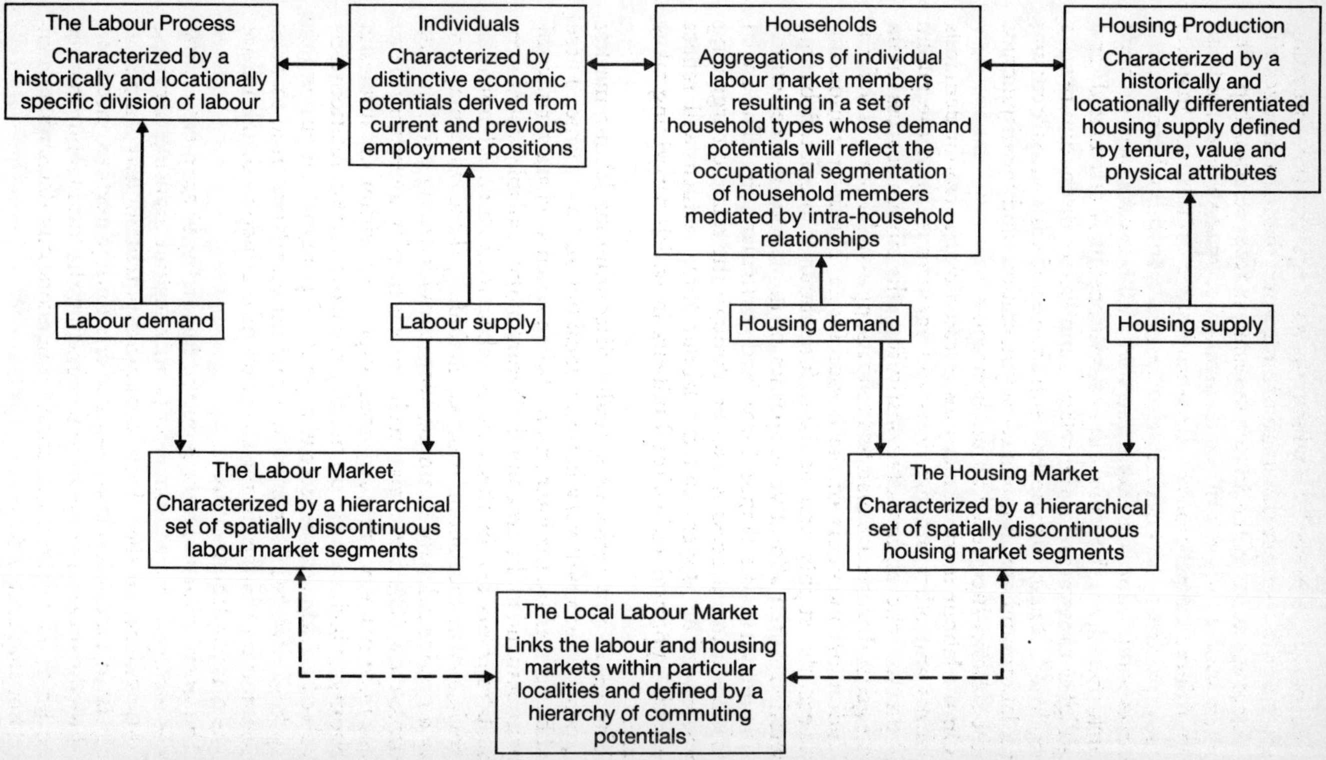

Figure 1.1 A framework for the analysis of housing market – labour market relationships.

On the other side of the relationship, the various structures of provision through which housing is produced constitute the supply of housing opportunities in any particular locality. This comprises both the historically existing and currently produced housing stock. The tenure characteristics of this supply play a major role in structuring housing market segmentation: first by defining the primary allocation criteria through which households access these opportunities, and, second, in determining the underlying value profiles of the stock through the differential impact of the various fiscal and financial structures by which demand is translated into price in each tenure. However, while tenure will represent a major line of demarcation, it is by no means the only one, and a variety of intra-tenure submarkets are likely to be present. The various physical characteristics of the stock – size, amenity and particularly location – also form major elements in defining the characteristics of housing supply, especially in value terms.

Interacting with this pattern of housing opportunity will be the range of households who constitute the demand side of the housing market. Households, which may be viewed as 'consumption units' in the housing market, are constituted by varying aggregations of individual labour market members, including those not directly involved in the formal labour market. Household demand capacities will principally derive from the position in the formal labour market of individual household members, but, as I argued above, these can be significantly modified by economic, social and cultural relationships within the household itself.

The outcome of this interrelationship is a locationally and historically specific set of spatially discontinuous housing market segments, reflecting both the structure of housing opportunities in the locality and the characteristics of households consuming these opportunities. The actual value profiles – prices and rents – of the housing market in any area represent the outcome of the interrelationship between household demand potentials, derived from, but not directly corresponding to, the labour market position of individual household members, and the underlying value structures of the tenure-based submarkets in the area.

Households are differentially 'sifted and sorted' by the allocative mechanisms of the housing market into the housing spaces created by the housing production process. Again, a dynamic element is injected into the housing market process by the relative bargaining positions of households in relation to the constraints imposed by the housing system. This is the arena of housing consumption struggles, both between different segments (for example, in certain inner cities between high income owners and low income renters) and between households and the suppliers of housing services.

And within segments, individuals (or at least their households) are forced into competition with each other for preferred locations and properties, in a similar way to which individuals compete in the labour market for preferred jobs. Indeed, the crucial role of both individuals and groups to actively mediate the outcome of these processes, through the bargaining processes which are integral to both labour and housing markets, must be fully recognized.

The schema also highlights the way in which the aggregation of individual labour market members into households represents a major disjuncture in the interrelationship between labour and housing markets. Most importantly, it is through this nexus that the two markets are functionally linked. The almost infinite variety of households and individual housing consumption potentials which results bears no direct relation to the position in the labour market of individual household members. Earlier rounds of capital accumulation within the household, achieved either through the ownership of domestic property or from other sources outside the formal labour market, contribute to the economic potential of individual households and add additional complexity to the relationship. The household also plays a central role in shaping behaviour, attitudes and perceptions to both work and housing and acts as the repository of social and cultural attitudes which influence the way in which housing is perceived and utilized. These behavioural characteristics feed through the market relationships through their influence on housing values.

It is also important to stress that the relationships between the various elements in the framework are dynamic in nature, slowly changing as a cumulative result of changes in the labour process leading to differential job loss or gain, changes in the nature and rate of dwelling production and demolition, and in the structure of domestic relationships and household characteristics. Thus the 'stocks' in each sphere are continually changing, although not necessarily in any systematic relationship with each other. Changes in any one component will nevertheless have repercussions throughout the system. Indeed, the linkages between the various elements in the schema are viewed as being fully reciprocal – there are no determined, uni-causal relationships.

Finally, in spatial terms, housing and labour markets within a particular locality are linked through the 'local labour market' which is largely defined by the commuting propensities of the various labour market segments which constitute it at any one time. These may be rather loose, as in the case of the commuting hinterland of a major metropolitan area, or much more tightly defined, such as that of a small freestanding town with a major single employer.

Concluding comments

This chapter has not set out to answer questions. Rather, it has sought to stimulate them. The objective has been speculative, and the aim has been to encourage discussion rather than to set up definitive 'models' of what is clearly a highly complex set of relationships. In particular, no attempt has been made to specify a typology of segmentation or to apply the discussion to any particular situation. Any validation of the concepts discussed could only emerge from their application in particular empirical studies. Consequently, it does not seem particularly appropriate to suggest conclusions *per se*.

Despite its speculative nature, however, I hope the chapter has gone some way towards shifting the nature of the prevailing debate on housing market–labour market relationships, away from the conceptually limited approach which starts with aggregate 'black box' empirical juxtapositions – models of regional house prices and incomes, for example – and towards an approach that would tease out the relationships involved on both sides of the equation, but focusing on the level of individuals and households in the process.

This is not an attempt to invalidate the macro-empirical approach. Rather, it would act as an additional perspective both to aid the specification of empirical approaches (for example, it raises questions of how you include the household–individual relationship correctly into the equation) and provides an additional level of understanding of the relationships involved to aid the interpretation of aggregate level approaches. Neither is it anti-empirical. The value of the framework suggested above is seen as much in terms of its role in structuring empirical analysis as in developing conceptual discussion.

However, it is argued that the application of spatial discontinuity theory to conceptualizing the structure of housing and labour markets and their interrelationships is an attempt to define both the divisions in these markets and the nature of the links between them in something more than an empirical, 'data determined', way.

The above discussion also stresses a number of additional, and I would argue, essential points. First, the links between housing and labour markets are only partially predictable because of the mediating influence of the household. At a conceptual, as well as an empirical, level, we need therefore to shift the analytical focus onto the household and its structure as the unit of analysis, rather than, or at least in addition to, the individual. For example, how do the various economic relationships within households affect housing consumption potentials? How do you allow for multi- and single-income households and informal economic household activity in analyses of labour market, housing market relationships? How do behavioural and other non-economic processes affect both housing consumption and labour market participation levels? These sort of questions need to be asked in order to develop our understanding of how labour and housing markets interact at the aggregate level.

Second, the degree of autonomy in the way in which both markets operate has major ramifications in understanding the relationships between them, as do the different scales of spatial resolution at which they appear to operate and the differential dynamism they exhibit. In simple terms, each market has a mind of its own, and changes in one are only partly related to changes in the other. These independent influences need also to be understood if we are to make any sense of the relationship between the two markets.

Third, an approach adopting concepts of spatial discontinuity theory would, hopefully, allow greater opportunity to expose the full range of processes and influences which structure housing and labour markets and the outcome of the interactions between them.

Perhaps the real world is simply too complex to make such an approach practical in other than limited locality-based studies. But it is argued that at the very least, the concepts discussed above may help to open up the 'black box' for a more perceptive approach to understanding the complex structures of housing markets, labour markets, the households and individuals which constitute them, the forces which shape them, and the relationships between them.

References

Agnew, J. (1981), 'Homeownership and the capitalist social order' in M. Dear and A. J. Scott (eds), *Urbanisation and Urban Planning in Capitalist Society* (London, Methuen, pp. 457–80).

Allen, J. and McDowell, L. (1989), *Landlords and Property* (Cambridge, Cambridge University Press).

Anderson, J., Duncan, S. and Hudson, R., (eds) (1983), *Redundant Spaces in Cities and Regions* (London, Academic Press).

Ball, M. (1982), 'Housing provision and the economic crisis', *Capital and Class*, 17, pp. 60–77.

Berger, S. and Piore, M. J. (1981), *Dualism and Discontinuity in Industrial Society* (Cambridge, Cambridge University Press).

Bosanquet, N. and Doeringer, P. (1973), 'Is there a dual labour market in Great Britain?', *Economic Journal*, 83, pp. 421–35.

Bowlby, S., Foord, J. and McDowell, L. (1985), 'For "love" not money: gender relations in local areas'. Paper presented at the Fifth Urban Change and Conflict Conference: *Industrial Restructuring, Social Change and the Locality*, University of Sussex, 16–19 April.

Burney, E. (1967), *Housing on Trial*, London (Oxford, Oxford University Press for the Institute of Race Relations).

Byrne, D. and Parsons, D. (1983), 'The state and the reserve army: the management of class relations in space', in J. Anderson, S. Duncan and R. Hudson, *Redundant Spaces in Cities and Regions* (London, Academic Press).

Clapham, D. and Kintrea, K. (1986), 'Rationing, choice and constraint: the allocation of public housing in Glasgow', *Journal of Social Policy*, 15, 1, pp. 51–67.

Community Development Project (CDP) (1976), *Whatever Happened to Council Housing*? (London, CDP).

Cooke, P. (1983), 'Labour market discontinuity and spatial development', *Progress in Human Geography*, 7, 4, pp. 543–65.

Cross, M. (1983), 'Racialised poverty and reservation ideology: blacks and the urban labour market'. Paper presented at the Fourth Urban Change and Conflict Conference, Essex University.

Doeringer, P. and Piore, M. (1971), *Internal Labour and Manpower Analysis* (Lexington, D. C. Heath).

Donnison, D. (1985), 'The rebirth of the domestic system', *Town and Country Planning*, April, pp. 120–2.

Edwards, R., Reich, M. and Gordon, D. (eds) (1975), *Labour Market Segmentation* (Lexington, Mass., Lexington Books).

English, J. (1982), 'Must council housing become welfare housing?', *Housing Review*, Sept/Oct, pp. 145–56, and Nov/Dec 1983.

Evans, C. and Cooke, P. (1985), 'Women homeworkers and the defeminisation of industry'. Paper presented at the Fifth Urban Change and Conflict Conference: *Industrial Restructuring, Social Change and the Locality*, University of Sussex 16–19 April.

Flett, H. (1977), 'Council housing and the location of ethnic minorities', *Working Papers on Ethnic Relations No. 5*, SSRC Research Unit on Ethnic Relations.

Foote, N. N., Abu-Lughod, J., Foley, M. M. and Winnick, L. (1960), *Housing Choices and Constraints* (New York, McGraw Hill).

Forrest, R. (1987), 'Spatial mobility, tenure mobility, and emerging social divisions in the UK housing market', *Environment and Planning A*, 19, pp. 1611–30.

Forrest, R. and Murie, A. (1983), 'Residualisation and council housing: aspects of the changing social relations of housing tenure', *Journal of Social Policy*, 12, 4, pp. 453–68.

Forrest, R. and Murie, A. (1986), 'Marginalisation and subsidised individualism: the sale of council housing in the restructuring of the British

welfare state', *International Journal of Urban and Regional Research*, 10, 1, pp. 46–65.
Forrest, R. and Murie, A. (1987), 'The affluent homeowner: labour–market position and the shaping of housing histories', in N. Thrift and P. Williams, *Class and Space* (London, Routledge and Kegan Paul).
Friedman, A. (1977), *Industry and Labour* (London, Macmillan).

Gershuny, J. (1978), *After Industrial Society?* (London, Macmillan).
Gershuny, J. (1983), *Social Innovation and the Division of Labour* (Oxford, Oxford University Press).
Gordon, D. M. (1972), *Theories of Poverty and Underemployment* (Lexington, Mass, Lexington Books).
Gordon, D. M., Edwards, R. and Reich, M. (1982), *Segmented Work, Divided Workers* (Cambridge, Cambridge University Press).
Greater London Council (1985), *The London Industrial Strategy*.

Haddon, R. F. (1970), 'A minority in a welfare state society: the location of West Indians in the London housing market', *New Atlantis*, 2, pp. 80–133.
Hakim, C. (1982), 'Homeworking in the London clothing industry', *Employment Gazette*, 90, pp. 369–76.
Hamnett, C. (1984), 'Housing the two nations: socio-tenurial polarisation in England and Wales, 1961–81', *Urban Studies*, 43, pp. 389–405.
Hamnett, C. and Randolph, W. (1986a), 'Labour market restructuring in Greater London 1971–81: Evidence from the Longitudinal Survey', City University Social Statistics Research Unit.
Hamnett, C. and Randolph, W. (1986b), 'The Role of Labour and Housing Markets in the production of geographic variations in social stratification', in K. Hoggart and E. Koffman, (1986) *Politics, Geography and Social Stratification* (London, Croom Helm), pp. 213–46.
Hamnett, C. and Randolph, B. (1988a), 'Socio-tenurial polarisation in London: a longitudinal analysis', *Urban Studies*, pp. 25, 338–98.
Hamnett, C. and Randolph, B. (1988b), *Cities, Housing and Profits: The Flat Break-up Market in London* (London, Hutchinson).
Harrison, P. (1983), *Inside the Inner City* (London, Penguin).
Henderson, J. and Karn, V. (1984), 'Race, class and the allocation of public housing in Britain', *Urban Studies*, 21, pp. 115–28.

Johnston, R. J. (1971), *Urban Residential Patterns* (London, G. Bell and Son).

Karn, V. (1985), 'The financing of owner occupation and its impact on ethnic minorities', *New Community*, pp. 49–63.
Kreckel, R. (1980), 'Unequal opportunity structure and labour market segmentation', *Sociology*, 14, 4, pp. 525–49.

Ley, O. (1981), 'Inner city revitalisation in Canada: a Vancouver case study', *Canadian Geographer*, 25, 2, pp. 24–45.

MacKensie, S. and Rose, D. (1983), 'Industrial change: the domestic economy and home life', in J. Anderson *et al.* (eds), *Redundant Spaces* (London, Academic Press), pp. 155-200

Malpass, P. (1983), 'Residualisation and the restructuring of housing tenure', *Housing Review*, March, pp. 44–5.

Massey, D. (1984), *Spatial Divisions of Labour: Social Structures and the Geography of Production* (London, Macmillan).

Massey, D. and Meegan, R. (1982), *The Anatomy of Job Loss* (London, Methuen).

Munro, M. (1986), 'Housing and labour market interactions: a review', Centre for Housing Research, Discussion Paper 12, Glasgow University.

Murie, A. (1983), *Housing, Inequality and Deprivation* (London, Heinemann).

Pahl, R. E. (1980), 'Employment, work and the domestic division of labour', *International Journal of Urban and Regional Research*, 4(1), pp. 1–20.

Pahl, R.E. (1984), *Divisions of Labour* (London, Basil Blackwell).

Pahl, R. E. (1985), 'The restructuring of capital, the local political economy and household work strategies', in D. Gregory and J. Urry (eds), *Social Relations and Spacial Structures* (London, Macmillan).

Parker, J. and Dugmore, K. (1976), 'Colour and the allocation of GLC housing', *Research Report No. 21*, Greater London Council.

Piore, M. J. (1982), 'American labour and the industrial crisis', *Challenge*, 25, 1, pp. 5–12.

Pratt, G. (1981), 'The house as an expression of social worlds', in J. Duncan (ed.), *Housing and Identity* (London, Croom Helm), pp. 135–80.

Pratt, G. (1982), 'Class analysis and urban domestic property. A critical examination', *International Journal of Urban and Regional Research*, 6, 4, pp. 481–501.

Randolph, W. (1987), 'Social change and residential polarisation in London 1961–1981', PhD thesis, London University.

Redclift, N. and Mingione, E. (eds) (1984), *Beyond Employment* (Oxford, Basil Blackwell).

Rex, J. and Moore, R. (1967), *Race, Community and Conflict: a study of Sparkbrook* (London, Oxford University Press).

Rex, J. and Tomlinson, S. (1978), *Colonial Immigrants in a British City: A Class Analysis* (London, Routledge and Kegan Paul).

Sassen-Koob, S. (1985), 'Capital mobility and labour migration: their expression in core cities', in Timberlake (ed.), *Urbanisation in the World Economy* (New York, Academic Press).

Saunders, P. (1982), 'Beyond housing classes: the sociological significance of private property rights in means of consumption', *Urban and Regional Studies Working Paper*, 33, University of Sussex.

Saunders, P. and Williams, P. (1984), 'Charity begins at home: some thoughts on recent and future developments in Urban Studies'. Paper presented to the British Sociological Association Sociology and Environment Group, November.

Shevsky, E. and Bell, W. (1955), *Social Area Analysis* (Stanford, Stanford University Press).

Simmie, J. (1983), 'Beyond the industrial city', *Journal of the American Planning Association*, Winter.

Smith, N. and Williams, P. (eds) (1986), *Gentrifications of the City*, (London, Allen and Unwin).

Storper, M. and Walker, R. (1983), 'The theory of labour and the theory of location', *International Journal of Urban and Regional Research*, 8, 1, pp. 1–41.

Taylor, P. J. (1979), '"Difficult to let", "difficult to live in", and sometimes "difficult to get out of": an essay on the provision of council housing, with special reference to Killingworth', *Environment and Planning A*, vol. 11.

Twine, F. and Williams, N. I. (1983), 'Social segregation and public sector housing: a case study', *Transactions of the Institute of British Geographers*, N.S. 8, pp. 253–66.

Ward, R. (1982), 'Race, housing and wealth', *New Community*, 11, pp. 3–15.

Watson, S. (1985), 'Shifting dichotomies – the restructuring of productive and reproductive relations'. Paper presented at the Conference on Labour and Housing Market Change, Parsifal College, London, December.

Williams, P. (1986), 'Social relations, residential segregation and the home', in K. Hoggart, and E. Kofman, (eds) *Politics, Geography and Social Stratification* (London, Croom Helm), pp. 247–73.

PART TWO

Understanding the connections

Introduction

[illegible]

Introduction

In this section we set out some of the possible directions that the study of labour and housing markets and their interrelations can take. As we saw in the first section, many of the discussions to date on the connections between labour and housing markets have been characteristically undertheorized. In that sense, Randolph's chapter represents a welcome attempt to come to theoretical grips with the divisions within the two markets and the links between them. In particular, as Randolph suggests, his chapter is intended to open up the discussion, rather than to foreclose it around the concepts of segmentation and discontinuity theory. In this section, each chapter has something to contribute to the shape of that discussion. In their own way, each chapter throws light on a different kind of connection that can be drawn between the two markets. However, whereas Randolph's theoretical framework is abstract and open-ended, the chapters in this section take a more concrete focus, illustrating theoretical insights where appropriate with evidence from the UK, from Australia and from the USA.

Forrest and Murie, for example, as well as Salt, examine the connections between housing and labour mobility, although they do so in quite different ways with varying results. Forrest and Murie explore the extent to which housing histories are shaped by career histories and vice versa, through a case study analysis of two contrasting localities in Bristol, one skilled working class, the other middle class. Salt, in contrast, examines the relationship between housing tenure and labour mobility to construct an argument about the pattern of labour migration within the UK. Marcuse's claims are of a different order. He argues for a greater emphasis upon class-based work practices to unravel the characteristics of the US housing market, especially those of gentrification and homelessness. Arguing against the view that cities are becoming polarized along tenure lines, he stresses the notion of the 'quartered city': a four-class pattern that cuts across tenure lines. Finally, Watson draws our attention to the ways in which a largely unacknowledged, male conception of paid work has distorted analyses of jobs and housing. Drawing upon Australian materials, she wants to take the analysis beyond the formal operation of the two markets by re-working the relation between homes and work to reflect the experiences of women. In fact, it is a characteristic of each of the contributors in this section that they wish to shift the focus of inquiry away from the formal operation of both labour and housing markets. Nonetheless, it is important to recognize that the 'markets' remain the benchmark from which each of the chapters start. Let us spell this out more clearly for each of the contributions.

For Forrest and Murie, and also for Salt, their understanding of the connection in market terms is clearly evident. Both chapters address the relationship between jobs and housing through the formal market mechanisms of the exchange process. On the one side, the labour market is considered primarily through the nexus of waged employment and, on the other side, the housing market is approached through the mechanisms of supply and demand for residential property. Neither account is exhausted by this focus and both chapters, as we shall see, go on to develop interesting and innovative conceptions of the relationship between jobs and housing.

Marcuse too, in his account of the interrelations between work and housing takes his starting point from market mechanisms but very quickly moves away from an arena in which buyers and sellers are the main players. As noted above, he wants to conceive the connection between work and housing provision in class terms, rather than in market terms. Marcuse's argument is an attempt to shift the focus of inquiry from (what he sees as) a rather limited, market view of the links between labour and housing towards an understanding of how various groups of paid workers (decision-makers, middle managers, cultural workers, the self-employed, and so forth) and their associated lifestyles are broadly connected with certain types of housing. The connection he is searching for is between the kinds of paid work that people do and how they use housing. The use and the role that housing fulfils for different groups of workers is a key issue.

Loft living in New York, for example, or the conversion of warehouse space into residential accommodation in the Docklands of east London, is seen by Marcuse as a response to the needs and aspirations of a class of professional and managerial workers who, although well paid and involved in creative kinds of work, are essentially unfulfilled by work. Gentrification of property in this context is one way of sublimating the desires and motivations of this class. And as the nature of work changes for this class then so too will the role that housing presently performs for them. The work that people do then and not how much they are paid for it is seen as an influential factor that changes the demand and indirectly the supply of housing. Class practices and tastes, translated from the kinds of work peculiar to the different classes, are thus considered to be far more significant than the commodity character of either wage labour or housing.

Watson's aim, as indicated above, is also to shift the focus of inquiry away from the formal operation of housing and labour markets, although in a direction that is quite different from the one that Marcuse takes. She wants to think of the relation between jobs and housing, work and the home, in gender terms. Work, she argues, takes its meaning from an association with male (formal) paid employment outside of the home, whereas work within the home is seen as unproductive and associated with women's unpaid domestic labour. This set of associations, she argues, is not accidental or arbitrary; they take their meaning from gender ideologies and serve to obscure the different kinds of work that women do – in the formal labour market, in the informal economy, and in the household.

To introduce a richer notion of women's work into the discussion of the links between home and work, Watson sets out a case for re-thinking the relations between the two sides; one that recognizes the shifting boundaries between the two. Drawing upon examples from Australia, many of which are resonant with trends in the UK and the United States, she points to real changes that bring into question received notions of the home–work boundary. For example, the recommodification of domestic labour such as childcare, laundry and cleaning, has shifted work out of the homes of middle-class women into the homes of working-class women. There is a shift here not only between paid and unpaid work, or between the public and private domain, there is also a shift in how work and housing may be conceived. Service work of this kind can be organized in a variety of ways – on a commercial basis, on a private basis within the household, or even through the state. Interestingly, the above example also raises another dimension alongside gender, that of class. It is middle-class households who are employing working-class women to carry out their domestic services. The implication here is that the meanings of home and work are likely to be different for the two classes and the women involved.

Thinking through the connections between work and housing, whether in class terms or gender terms, or indeed through a meshing of the two, does indeed take us beyond market relations, often into areas and issues that previously have remained outside of the debate on the housing–labour market connection. While the legitimacy of these areas cannot be contested, they have yet to displace the significance of markets as a central focus. Turning again to the chapters by Forrest and Murie, and by Salt, it is apparent fairly early in their respective accounts that, even though they share an interest in the structure of housing markets and how they are shaped by career histories and mobility within the labour market, they are nonetheless working within quite different explanatory frameworks. This then begs the question of what kind of market relations they are talking about. And, what is more, whether the two interpretations are compatible.

There are two points worth pulling out here. The first is that Forrest and Murie examine each side of the housing market–labour market relation in a way that is significantly different from Salt. Forrest and Murie are concerned to show that there is no simple relationship between housing tenure and labour market position. Indeed, they go to considerable lengths to demonstrate the divergence of housing histories within one tenure, owner occupation. By taking a group of households at the top and bottom ends of the owner-occupied market they are able to show that the housing histories of the two groups were shaped in qualitatively different ways by job characteristics and job histories (of the man's employment). Salt arrives at a similar conclusion about the direction of causality, in that he too stresses the role of male career histories in attempts to explain patterns of labour and housing mobility. Also, both chapters emphasize the highly segmented nature of the labour market and the different housing trajectories experienced by those workers outside of and those workers in corporate internal labour markets. The type of job and even the type of company, its recruitment and retention policies, are all seen as significant factors behind the variation in residential mobility rates. But here the similarities start to shade out and the differences in analysis take hold.

Starting with the housing side, for Salt the concept of tenure is treated as largely unproblematic; as a kind of limiting factor upon housing mobility, with only marginal effect upon labour migration patterns. Housing is interpreted mainly in terms of people's tenure position and regarded as one type of barrier to labour mobility among other types. So, taken together, the Rent Acts and the decline of the private rented housing market, the inflexible allocation of public sector housing, and the geographical distortion of house prices in the owner-occupied sector, are seen in different ways to have raised obstacles to labour mobility. Where Forrest and Murie would take issue with this account is that tenures are represented as a unified and generalizable experience for the households within them. Forrest and Murie advance the view that tenure divisions may not be the major line of divide within housing; that differences *within* home ownership may be a more significant determinant of housing opportunity than differences *between* housing tenures. Moreover, they argue that the divergent opportunities of households within the same tenure are related to labour market segmentation – differing migration and mobility patterns are related to the type of career held and the type of company a person is employed by. Housing tenure, then, in their view, is something that should not be taken for granted; it too needs to be re-examined, before an examination of the connections with jobs and labour markets can be carried out.

On the employment side, the approaches of Forrest and Murie, and of Salt exhibit greater similarity. Both stress that labour market divisions and, in turn, residential mobility rates, are about more than income differences, variations in job security or status differences between jobs. It is the type of job that counts, as noted above. What is meant exactly by type of job differs between the two chapters however. For Salt, a job is defined as a discrete set of tasks associated with a particular occupation, whereas for Forrest and Murie certain occupations are associated with broader 'ways of life'. There is an experiential dimension included in their analysis which contrasts with Salt's more formal behavioural analysis. Indeed, Forrest and Murie's emphasis upon occupational lifestyles connects with Marcuse's view that different occupations create a set of needs and aspirations which can, in part, be realized through different uses of housing. That having been said, it is unlikely that Forrest and Murie would share Marcuse's viewpoint, especially as their respective notions of class and class lifestyles differs substantively. Nevertheless, the difference in approach between Forrest and Murie, on the one hand, and Salt, on the other, remains. And a striking illustration of this contrast can be gleaned from Forrest and Murie's reference to professional households (not tenures) which are locked into global as opposed to local space, with housing histories and career histories that span national and even international internal labour markets. There is an entirely different way of life involved here; one that contrasts markedly with the kinds of work associated with attachment to local networks and local markets. And the contrast is not solely one of class, as Forrest and Murie are at pains to point out.

So, it would appear that the links between jobs and housing are certainly complex, and even the nature of each side of the connection is mediated rather than unmediated. This brings us to the second point concerning the differences in the explanatory frameworks of Forrest and Murie, and of Salt, namely what methods they adopt to handle complexity – what kinds of evidence they use and how they generalize from their research findings. At a glance, the obvious differences between the two is that the former attempt to make sense of complexity by sifting individual housing and work experiences through the use of qualitative in-depth research techniques, whereas Salt uses quantitative research methods to reveal something of the complexity. Perhaps more important, however, is that Forrest and Murie are not attempting to generalize about two groups of households in the owner-occupied market, but rather they are trying to identify the key features which shape their housing histories. The distinction between isolating the causes behind trends and generalizing about the trends themselves is significant. It highlights one of the main reasons for adopting either qualitative or quantitative research methods. According to Forrest and Murie, the use of case studies enables them to expose more subtle forms of connection between labour and housing markets that perhaps more statistically-based, aggregate trends would not reveal. That is not to say that they consider qualitative approaches as in some sense better than quantitative approaches. Rather, the two kinds of methods have different jobs to do. In that sense, the two approaches can be seen as complementary.

If that is the case, then a further question arises; namely, is the approach of Salt compatible with that of Forrest and Murie? Certainly the two accounts cannot be added together in any simple manner, as if a case study approach were the natural precursor of a more standardized approach. Yet both chapters do suggest that the reasons for varying mobility and migration patterns are to be found in the operation of labour markets and the processes that have led to their segmentation. How they arrive at that suggestion, however, the kinds of reasoning that they employ, are not entirely complementary – as we have seen. When Forrest and Murie talk of causal connections between labour and housing markets they do so after having taken apart both markets in a way that is different from Salt. How the two accounts build the connections between the two markets therefore is quite distinct.

The issue of causality is of crucial significance to Marcuse also. If Forrest and Murie tread warily among questions of tenure, class and labour market divisions, Marcuse exhibits fewer doubts in his analysis. Central to his account, as indicated earlier, is the view that the actual connections between jobs and housing varies by class position. And in what he terms the 'quartered city', four classes – the upper class, the professional/managerial class, the working class, and the very poor class – are active in shaping and influencing housing divisions. He, together with Forrest and Murie, is keen to play down the importance that, hitherto, has been attached to the links between housing tenure divisions and labour market cleavages. Unlike Forrest and Murie, however, who wish to qualify class-based connections between home and work, Marcuse is keen to establish an *internal* relationship between class-based work practices and housing.

What does it mean to refer to internal relations between jobs and housing? What kind of causal connections are involved? For Marcuse, the distinction between internal and external relations is relatively straightforward. Internal relationships between jobs and housing hold primarily between different kinds of work and certain uses of housing. An example of this kind of relation was referred to earlier, namely the link drawn by Marcuse between gentrified housing and the professional and managerial class, those who held aspirations to convert property into a certain style to complement their professional activities. In contrast, external relations between jobs and housing are said to hold in the market-place, where the demand preferences of the four classes are met or not, as the case may be. More broadly, with an internal relation, the relation itself is likely to be a kind of structural relation in which the two sides combine in such a way that the relation itself shapes the whole. With an external relation the connection between the two sides is contingent and therefore the content of the relation may vary considerably over time and across space. Put more simply, with an external relationship, one side does not depend on the other to the same extent as the sides of an internal relation.

In Marcuse's version of the internal–external relation the distinction is not drawn quite as sharply. Yet, it is apparent that he is using the distinction to convey something of the *prior* difference between the variety of forms that work and housing can take, and how such differences go on to take their character from what happens in the markets. There is a temporal distinction in play here, as well as a shift in the level of abstraction between the two kinds of relation. For Marcuse, the internal relation between jobs and housing takes precedence, with a clear causal direction from changes in the kinds of work done to changes in the form of housing. And, in his view, it is these changes that are shaping the 'quartered city' in the USA; a reflection on the ground as it were of changing class practices.

Thinking through the relationship between jobs and housing as one that may take a variety of forms is also common to Watson's approach and, like Marcuse, owes much to the work of Pahl. In his book *Divisions of Labour* (1984), Pahl focuses on the household as the major agent involved in different kinds of work, within and outside of the formal economy, and within the home. Different household structures (for example, high income, multiple earner households or single parent households) are involved in different kinds of work and employ different kinds of labour to meet their respective household needs. Watson, as noted earlier, takes this analysis further to show how gender ideologies have limited the analysis of work to the formal domain of paid male employment. Career paths in, for example, the accounts given by Salt and by Forrest and Murie are male career paths, and in both cases housing histories and housing movements are shaped by the dictates of the man's employment. Forrest and Murie recognize this practice of gender myopia, although it is left to Watson to show how the ideological assumptions that underpin this practice influences the kind of questions that researchers ask about the structure of labour and housing markets and, perhaps more crucially, where they seek the causal connections between the two.

Gender divisions, for Watson, do not only cut across class lines and household types, they also cut through many of our dominant conceptions of what kinds of work and what kinds of housing are important. Thinking about these conceptions, rather than thinking with them, is what this section is all about. This is not to suggest that the most fruitful path of analysis is to shift our conceptions beyond those of the formal operation of markets. We have yet to exhaust our understanding of the formal connections that link housing and labour markets. Rather it is to serve as a reminder that the formal boundaries drawn around the two markets and indeed some of the formal links drawn between them such as the house price mechanism and the wage mechanism do not exhaust the links between jobs and housing.

References

Pahl, R. E. (1984) *Divisions of Labour* (Oxford, Basil Blackwell).

2 *Housing markets, labour markets and housing histories*

RAY FORREST & ALAN MURIE

Introduction

The main aim of this chapter is to offer a contribution to the discussion of methods and perspectives which are useful in developing a better understanding of housing and labour market links. In doing this it is essential to acknowledge that both housing and labour markets are changing and the links between these markets are affected by these changes. For example, one perspective on the relationship between housing and labour markets is that the production of housing has responded to patterns of demand which reflect employment and labour market trends; what housing has been built and where is linked to patterns of industrial development. However, as the housing market has matured the role of second-hand properties has become more significant. Changes in employment and housing demand are less obviously reflected in housing production and involve a more complex range of market responses. Even where market responses are very dramatic – in Aberdeen or Corby – the interaction with a mature housing market requires a wider analysis than is needed for histories of what is built and where.

Tenure status and place of residence are important indicators of social status and wealth. In many circumstances they are the most obvious clues to social and class position. But in the literature on social stratification, housing position is generally treated as a correlate of occupational and economic status. Recent contributions have included attempts to move beyond this perspective and to suggest that housing is a pivotal differentiator of consumption status and an important factor in political and voting behaviour (Dunleavy, 1980; Saunders, 1984). Indeed, at the extremes there are stances which suggest either that housing has some independent, major importance in social inequality and stratification, or that housing simply mirrors employment position and income, and has no separate impact on a social system in which position in relation to production is paramount. A problem which remains in the debate is a tendency to generalize about the relationship between and within tenures from static comparisons of groups at particular points – and a tendency to generalize about the owner-occupied market as a whole from particular experiences in particular locations. Recent trends in the housing market in London and the South-east figure prominently in perceptions. Arguably, the extraordinary levels of house price inflation which have been evident have contributed to an exaggerated view of the rewards which are on offer. While there are undoubtedly people whose housing wealth (measured in terms of the potential exchange value of the dwelling they are currently occupying) bears little relation to their earned income, this group may only represent an exotic minority (and be concentrated in areas of high house price inflation). In some cases, the value of residential property in the estates of deceased persons will seem quite disconnected from their employment histories and average lifetime earnings. In other cases, residential relocation or retirement from a high paid to a low paid area will enhance post employment lifestyles. For the majority, however, housing wealth is likely to be in tune with employment (or non-employment) histories.

Where housing research has referred to aspects of employment and labour markets, both methods and principal foci have tended to limit relevance to present concerns. The substantial body of social survey evidence accumulated over a long time period has tended to refer to employment status and occupationally based social class and where it has focused on residential mobility has usually been restricted to recent moves – in the two years prior to interview. Reference has conventionally been made to reasons for moving and to 'job', 'housing' and other factors influencing the decision to move and housing search behaviour. The perspectives drawn from this work have tended to be arid and depersonalized. And both social survey based research and approaches based on urban managerialist or political economy perspectives which did not favour survey methods (tainted with positivistic associations and an implied degree of individual autonomy and choice to the neglect of structures which constrain and determine life chances) have tended to produce snapshot views rather than emphasizing change or process. Knowledge of the operation of the housing market has remained largely synchronic rather than processual. Not only do the methods used in these surveys offer little prospect of illustrating the processes involved but the nature of questionnaire design and the framework of analysis are indicative of the limited interest in housing and labour markets and of the limited conceptualization of this relationship. Such conceptualization needs to be revised in the light of changing economic and housing environments.

One of the most obvious links between housing and the labour market is the process of labour mobility. The fact that employees moving for job reasons will often (especially if a longer distance move is involved) have to change where they live, means that there is an interaction between the two markets which cannot be ignored. The housing market and particular tenures in it have tended to be typified in terms of barriers to mobility. The conventional picture focuses on barriers associated with the supply of rented dwellings or access to it. In a market increasingly dominated by home ownership, this focus is being altered with an awareness of problems of access and supply in the owner-occupied sector and problems of differential house prices and rates of return. This shift in focus is particularly relevant because there has been a reduction in geographical mobility in Britain since the early 1970s and a decline in job changing and moving house (Davis, 1986). Higher status socio-economic groups account for a greater share of migration and the numbers of economically active persons in these groups has grown. A significant proportion of moves that are employment-related do not involve changes of employer but occur within internal labour markets. It is likely that these moves increasingly involve substantial fringe benefits directly related to house purchase and sale. In this context, contractual migration and company personnel and management development policies may be much more important determinants of housing behaviour than traditional studies of residential behaviour suggest. Employers view the costs of assistance with housing and migration in comparison with the value of training and experience. And such assistance can contribute to an increasing polarization in forms and processes of residential mobility and fuelling rapid trajectories through the owner-occupied market for those households in privileged employment positions.

In other areas considerable comment has been made about the 'company welfare state' and the various fringe benefits available to particular groups (for example Field, 1981; Green, Hadjimatheou and Smail, 1984), but there has been little systematic discussion of this dimension of the housing system. Forms of employer-related housing assistance have always been around (see, for example, Bateman and Burtenshaw, 1973), but they are likely to appear to be growing in importance for a number of reasons. First, home ownership has grown rapidly in recent years. There are more home owners now than there were, say, ten years ago and socio-tenurial polarization means that those in the occupational categories which are most likely to move between regions and for job reasons are more likely to be owner occupiers. The vast majority of employment-related moves involve moving from one owner-occupied house to another. In 1985, of the 32,000 employees relocated by their companies, 81 per cent were home owners (Merrill Lynch Relocation Management International Limited, 1986). And a more detailed study of managers and professional staff who had relocated in the last ten years found that over 90 per cent were home owners (IMS, 1987). Moving is an expensive business for home owners and employees may be reluctant to move unless their firm is prepared to pay the associated costs. House price differentials can be a significant dimension of this. The IMS study found that longer distance moves were more likely to be from north to south and that among employees who had refused a move, over a third cited higher house prices as an important discouraging factor (p. 65). A recent survey of employers in mid-Berkshire found that regional house price differentials were the major barrier to recruitment (Parsons, 1987).

Second, the development of employer assistance with housing costs, as with other fringe benefits, increases inequalities beyond those indicated by wages and salaries. Government has done nothing to restrict their growth and favourable tax treatment has contributed to their popularity. Remuneration in the form of fringe benefits is particularly attractive to higher wage employees and those liable to pay income tax at higher rates. Fringe benefits which generate higher levels of investment in housing are particularly attractive because of the relatively good performance of housing as an investment (albeit with a strong regional differentiation) and because of the potential for access to credit and realization of wealth through trading down or leakage) without capital gains or other tax complications. Home ownership occupies a prominent position in widening social divisions and forms of subsidy and assistance which enhance trajectories through the owner-occupied housing market are an important mechanism for employers. They take on particular significance when access to home ownership is impossible for some households and for others with a precarious foothold mortgage and where maintenance costs are a major burden.

Third, the growth of housing costs, rising real interest rates and widening house price differentials mean that costs associated with home ownership have grown more than other costs. The pressure of house prices on London salary weightings is an obvious example. Access to home ownership in London and the South-east is increasingly restricted. Given the widening gap in house prices between the north and the south of England, movement from one end of the country to the other requires increasingly generous forms of assistance from employers. This development not only contributes to a privileged circuit in the housing market but is a matter of increasing concern to employers. It involves aspects of recruitment, of preventing quitting, generating loyalty and facilitating future mobility. Part of this concern is reflected in the setting up of the Employee Relocation Council by the CBI. At its inaugural conference in June 1986 price differentials within the owner-occupied market were identified as a major factor affecting resistance to relocation among employees and the cost of relocation (CBI, 1986). It is somewhat ironic that home ownership should have come to be seen as the barrier to labour mobility, rather than council housing which has conventionally been 'blamed' for immobility. It is even more ironic that those so committed to the free market in other spheres should (at least implicitly) suggest the need for state intervention over house prices. The fact is that from an employer's view at least the problem of labour mobility is a problem for the employed not the unemployed. Generally, the unemployed have few opportunities on offer in other areas.

Long distance migration for job reasons, it seems, is more likely to be in BMWs rather than on bikes and it can be heavily subsidized by employers. Evidence on employer-related benefits is patchy. There is, for example, no systematic information on the number of households with access to low interest mortgages although some work has been carried out on assistance with moving costs (see, for example, Salt, 1985).

Households and housing histories: methodological observations

The research referred to in the rest of this chapter provides a basis to identify the importance of employer assistance with housing in the housing histories of affluent home owners and contributes to discussion about interaction between housing and labour markets in another way. There is a developing literature on the meaning of home ownership and the connections between tenure and class (e.g., Edel, 1982; Gray, 1982; Pratt, 1982; Forrest, 1984; Harloe, 1984; Saunders, 1984). At the core of this debate is an argument that key contemporary social divisions are between those who own, or are in the process of buying, their dwelling, and those who continue to rent. This debate, as with that around mobility, demands a greater depth of understanding of contemporary housing experiences. This has provoked interest in more qualitative in-depth sociological research not only to provide a different and more rounded account of housing provision but to generate new questions which might be pursued using larger scale survey techniques.

Until recently, much housing research has either ignored the labour market altogether or simply linked housing market position to occupational status in a way which ignored job history and career migration. Access to housing has been regarded as determined by income and price and the links between housing and labour markets have been seen in terms of these rather than any other dimensions. The value of individual housing histories is that they highlight the complexity of the links between the two spheres.

The work referred to in this chapter was designed to develop different perspectives and methods in housing research. It emphasized individual experience and related housing experience and housing histories to aspects of employment and job history. In particular, it was concerned with the extent to which housing histories are partially shaped by labour market position and vice versa, the extent to which residential migration or non-migration are associated with occupational careers, the role of occupational housing benefits, the distinction between career spiralists and those with strong locality attachment and little career potential, and the extent to which entrepreneurial energies were being channelled into housing decisions as suggested by Farmer and Barrell (1981). The remainder of this chapter outlines the method and results of this research and refers to its implication for discussion of the relationship between housing and labour markets.

Interviews with a group of households at the top and bottom ends of the owner-occupied market had a number of objectives. By taking two dissimilar areas – one working class with low/moderate house prices, and one middle class with high prices, the potential to highlight different housing and labour market histories was increased. The interviews were designed to provide a greater knowledge of social differentiation in home ownership but also to break away from traditional housing survey and provide different accounts of the experience of housing. These accounts were intended to be less led by preconceived categorization and emphasis and to reflect individuals' own accounts of how and why their housing situations had changed throughout their lives. The main research technique was to construct detailed housing histories over an individual's lifetime. While housing histories have been constructed by others they have tended to focus on the early stages on the life cycle, to emphasize recent residential movement (to the neglect of non-movement or longer term patterns) or to rely on a limited data set derived through precoded questionnaires carried out by market research agencies with limited opportunity for individuals to express their views and experiences in their own terms (see, for example, Holmans, 1981; Ineichen, 1981; Holme, 1985; Thorns, 1985). Having tested a less structured approach in a study of council tenants it was decided to move nearer 'purer' life histories and allow more room for subjects to speak for themselves. Nevertheless, a *totally* unstructured format was rejected and it was decided to collect a common set of key details on a short interview schedule which also provided the framework for a more free flowing taped discussion.

An approach which would maximize open-endedness and seek to eliminate interviewer bias altogether was rejected although much of the literature on biographic method would emphasize these qualities (see, for example, discussions in Bertaux, 1981 and Plummer, 1983). It was not the intention to produce oral histories but contextualized case studies – contextualized in the sense of being structured and influenced by argument and debates produced from entirely different perspectives and methods. While allowing room for individuals to speak for themselves, we nevertheless brought with us a knowledge of the operation of the housing market and a whole baggage of concepts. Another reason for combining a structured core of base data with relatively open discussion was to facilitate the analysis of the information. A small number of lengthy, unstructured interviews can produce a mass of impenetrable transcripts. There is little point in producing rich accounts which are unusable. It was preferable to sacrifice some of that richness for a basic framework around which other information could be integrated.

In micro-sociological work on individuals there is always a problem of selecting appropriate persons. Mitchell's (1983) strictures that reliability, extrapolation and generalization depend more on the validity of the analysis, knowledge of context and general theory than the representativeness of events are important in considering this material. Our concern in any event was not to produce typical or representative cases but illustrative cases which would be complementary to other research in which large numbers and representativeness were more prominent concerns. In this case we sought illustrative cases from home owners in two localities selected as high price and lower price areas. The two groups at the top and bottom ends of the market were occupying very similar dwellings in similar locations (within each locality) and thus had a shared relationship to the broader urban fabric (e.g., transport, shops, clubs, churches). Interviews were concentrated in a small number of streets in each area. Housing histories were traced from birth so that in the case of married or cohabiting couples two separate histories were constructed.

The localities

As is evident from Figure 2.1, the two localities are strikingly different visually and physically. The wealthy area (A) is in a leafy wooded suburban part of Bristol. Two of the roads included are of large detached houses in large gardens (one third of an acre or more). Most of these houses were built between the wars, many to high standards by a respected local builder. These are large houses with large rooms and five or more bedrooms. Most have been modernized to include new kitchens and two or more bathrooms and WCs. However, most retain some original features including wooden panelling on walls, original fireplaces and wooden parquet flooring. The wealthy locality also included a group of newly built properties on a site where development was still taking place. These are executive dwellings marketed at £80,000 or more (1985 prices). The whole area is regarded as select and as the highest status residential area in Bristol (for a more detailed discussion of the affluent area see Forrest and Murie, 1987).

In contrast, the working-class locality (B) is a densely built area of terraced housing developed at the turn of the century – some above old mine workings. It is a stable affluent working-class district close to the city centre and the old Wills tobacco factory which provided relatively well paid jobs until its recent closure. These are houses with small rear gardens and very small front gardens. They were built without internal bathrooms or toilets although these have mostly been added by rebuilding rear extensions. Rooms are small although in some cases two rooms have been knocked through into one. Some of the dwellings are built on steeply sloping land with steps up or down from street level to front doors or with changes of level within the dwelling – notably between the body of the house and the rear extension.

Tables 2.1 and 2.2 provide details of differences in house prices in the two areas. Differences in the occupational and other characteristics of residents are also detailed. Those in the wealthy area are professionals, executives and small business proprietors in contrast to the skilled and semi-skilled workers in the other area. The two areas conform in many respects to the descriptions of such areas contained in the housing literature. The working-class area is not a gentrified or abandoned district. Many residents have themselves lived most of their lives in the area and it is common for them to have relatives living in the immediate neighbourhood. It is generally regarded as a close knit community with strong kin and friendship links. The majority of those interviewed were 'locals' who had lived in or near the area throughout their lives or had left it, for relatively short periods only, during the War or

(A)

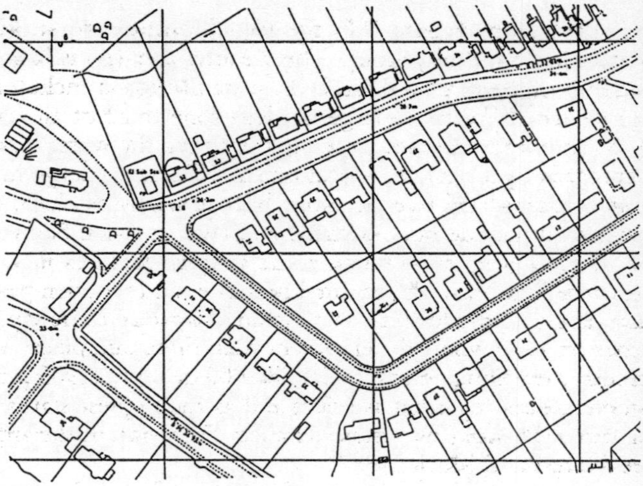

(B)

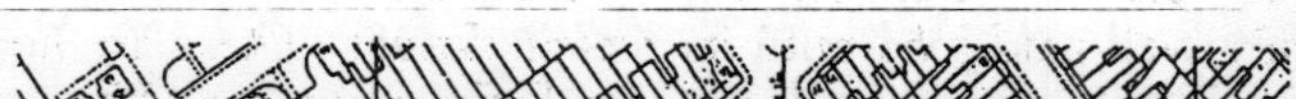

Figure 2.1 Contrasting images of home ownership

for jobs. This applies equally to males and females included in the study. One of the respondents had spent her whole life (sixty-six years) in the same dwelling and her husband had lived in only one other dwelling, some two streets away. One woman offered the following description of the locality:

> It's friendly and many people have relatives around here. Many people in this street, their families grow up and marry and buy houses just round the corner. There's quite a lot of that situation around here (authors' transcripts).

In contrast, those interviewed in the affluent area were not locals, had rarely been born in the neighbourhood, or indeed in Bristol, and had moved around the country usually in connection with their or their parents' job. They were unlikely to have relatives or friends living nearby or in Bristol and the locality was rarely regarded as either close knit or a community. There was a strong sense of social isolation which was particularly evident among working males and slightly less so among women outside paid employment with children and school and related activities which provided some contact with neighbours. But even here, catchment area changes had encouraged more private schooling so that the links between school and locality were tenuous and diverse. The males tended to refer to colleagues and acquaintances rather than friends. And social isolation through work was compounded through the form and layout of their dwellings and the use of and dependency on (usually) two cars. As one respondent remarked:

> . . . the only thing that connects these houses is the sewers and a touch of income. Everybody comes and goes in cars (authors' transcripts).

Another commented that he had not seen his neighbours since Christmas. (The interviews were in June.) And one couple told the (apocryphal?) story of neighbours they had met:

> . . . who said their daughter had just got married to somebody she had met at a northern university and it transpired that they'd both lived in this road and they'd both lived here all their lives and their parents had lived here all their lives. One at the top of the road and one at the bottom and they did not know one another. They had met by chance at university and finished up getting married. That's how close knit this community is! (authors' transcripts).

For some this was a positive feature of the area – high priced privacy. For others there was a certain sadness at the lack of neighbourliness. There was also a difference between patterns of contact in the older streets in the affluent area and the new executive development referred to earlier. In the latter, neighbours had been brought together in adversity when unwelcome changes were made to the layout and density of the development. Property values were threatened and a residents' association had been formed to combat the higher densities and buildings proposed.

> I would say that events over the last year have created a close knit community . . . I think it's fair to say that if this hadn't arisen, if we hadn't had these communal problems, I'm not sure that we would know our neighbours (authors' transcripts).

For another couple nearby buying on a new estate formed part of a strategy for coping with frequent and disruptive job moves and loss of friends:

> . . . if you buy a new house on an estate everybody's in the same boat and it doesn't matter whether they've moved a quarter of a mile or two hundred miles they are to some degree in a new area, in some degree strange to it and therefore you tend to get usually a community spirit within the estate and it's easier to make friends that way (authors' transcripts).

In some senses the localities split into three rather than two. And the problems of home ownership, of discontentment with design, quality, price and values were not to be found in the working-class area but in the new executive development where people were paying £80,000 or £90,000 for the locality rather than the dwelling and for a school catchment area which had subsequently been changed. They were also differentiated to some extent by stage in the family life cycle. Those in the new development were on similar forms of employment and on similar trajectories but at an earlier stage.

The housing histories

Tables 2.1 and 2.2 summarize the job histories of respondents into broad categories. Those in the affluent areas were more likely to have remained at school longer, to have higher education or professional qualifications and to have moved around the country or

Table 2.1 Profiles of home owners in two contrasting localities in Bristol, 1985

The affluent area

Interview number	Age		Current household type	Age left parental home	Age first house purchased	Age at marriage	No. of homes since parental home	No. of owner occupied homes	Price of first house purchased (£)	Value of current dwellings (£) (1987)
1	58	M	2 persons	23	30	30	8	2	4,000	80,000
2	54	F		17	26	26	10	2		
3	49	M	With adult	22	25	22	10*	6	2,300	125,000
4	47	F	family	20	23	20	10*	6		
5	58	M	2 persons	24	24	24	9*	6	350	80,000
6	51	F		17	17	17	9*	6		
7	59	M	With adult	23	23	23	4*	4	3,250	120,000
8	49	F	family	23	23	23	4*	4		
9	86	M	Single person	18	43	26	11	4	1,800	100,000
10	40	M	3 generations, including	25	25	25	3*	3	4,150	140,000
11	37	F	2 young children	17	22	22	5	3		
12	53	M	4 school age	18	30	29	18+	3	1,000	125,000
13	48	F	children	19	25	24	11+	3		
14	67	M	2 persons	21	35	32	8+	1	3,400	100,000
15	66	F		22	34	31	7	1		
16	61	M	2 persons	35	36	35	3*	2	3,650	110,000
17	50	F		17	25	24	4+	2		
18	56	M	2 persons	21	40	23	20+	2	7,700	100,000
19	57	F		22	41	24	20+	2		
20	45	M	3 school age	24	24	24	7	3	5,200	135,000
21	42	F	children	20	21	21	8	3		
22	42	M	1 young child	23	23	23	5*	4	4,600	150,000
23	41	F		22	22	22	5*	4		
24	47	M	2 persons	19	32	19	10	2	8,750	140,000
25	47	F		19	32	19	10+*	2		
26	41	M	1 young child	24	24	24	6	4	3,600	112,000
27	42	F		25	25	25	5*	4		
28	42	M	3 young	26	26	26	6*	6	4,995	110,000
29	39	F	children	23	23	23	6*	6		
30	34	M	2 young	19	31	31	6	2	19,000	105,000
31	34	F	children	18	22	22	4	2		
32	56	M	3 young	29	29*	29/43	7	5	29,500	85,000
33	40	F	children	21	30	28	5	2		
34	47	M	? young	24	33	33	10	3	7,800	80,000
35	40	F	children	20	26	26	7	3		

* Left home to marry

Notes: (a) No. 32 was the only person to have married twice and to have been a home owner prior to first marriage.
(b) In almost every case mothers had not been in paid employment.

Orientation to locality	Ever received housing subsidy from employer	Job history	Present occupation	Father's occupation
Migrants	No	Medical Career	Disabled	Bank Cashier
		Domestic	Housewife	Clergyman
Migrants	Yes	Executive Career	Senior Executive	Skilled Engineer
		Secretarial	Housewife	Club Secretary
Migrants	Yes	Executive Careeer	Senior Executive	Teacher
		Domestic	Housewife	Teacher
Neo-local	Yes	Exec./SE Bus.	SE Dealer	Estate Agent
		Secretarial	PT Secretary	Shopkeeper
Migrants	Yes	Banking Career	Retired	Plant Manager
		Exec., SE Bus.	SE Business	Farmer
Neo-local	No	Medical Career	Dep. Nursing Off.	Civil Servant
		Prof. Exec.	Senior Executive	Engineer
Migrants	Yes	Career		
		Domestic	Housewife	Civil Servant
		Medical Career	Semi-retired	Civil Servant
Migrants	No	Medical Career	Semi-retired	Jeweller
		Family Firm	SE Business	Insurance
Neo-local	No			Company
		Family Firm	PT Family Firm	Shop Manager
		Services Career	Senior Executive	Skilled Engineer
Migrants	Yes	Domestic	Housewife	Accountant
		Architect/Family	Family Firm	SE Ships
Neo-local	No	Firm		Chandler
		Secretarial/Family	Family Firm	Company
		Firm		Director
Migrants	Yes	Family Firm	Managing Director	Company Manager
		Domestic	Housewife	Contractor
Neo-local	Yes	Services Career	Bus. Consultant	Services
		Domestic	Housewife	Carpenter
Migrants	Yes	Banking Career	Regional Manager	Tool Maker
		Domestic	Housewife	Printing Manager
Migrants	Yes	Executive Career	Senior Executive	Draughtsman
		Education Career	P.T. Teacher	Engineering
Migrants	No	Self-made Family	S.E. Businessman	Carpenter
		Business		
		Medical Career	Housewife	Bus Driver
Migrants	Yes	Executive Career	Senior Executive	Printer
		Domestic	Housewife	Bank Manager
Migrants	Yes	Professional Career	Civil Servant	Clergyman
		Domestic	Housewife	Invalid

Table 2.2 Profiles of home owner in two contrasting localities in Bristol, 1985

The working-class area

Interview number	Age		Current household type	Age left parental home	Age first house purchased	Age at marriage	No. of homes since parental home	No. of owner occupied homes	Price of first house purchased (£)	Value of current dwellings (£) (1987)
1	65	F	Widow with adult son	23	25	23	1	1	800	18,000
2	30	F	Widow	20	19	20	1	1	8,300	20,000
3	42	M	2 school age	23	25	23	6	3	2,250	22,000
4	41	F	children	22	24	22	6	3		
5	83	M	2 persons	14	44	26	20+	4	400	22,000
6	79	F		16	40	22	18	4		
7	75	M	2 persons	28	35	28	2	1	550	22,000
8	67	F		19	27	19	2	1		
9	42	M	1 school age	20	30	22	7	1	5,000	23,000
10	45	F	child	18	33	25	2	1		
11	59	M	2 persons	22	34	22	5	1	800	21,000
12	56	F		19	31	19	5	1		
13	82	M	Single widower	25	53	21	3	1	360	22,000
14	70	M	2 persons	31	38	31	1	1	1,250	24,000
15	66	F		never left	34	27	0	1		
16	56	M	With adult	24	26	24	7	2	1,350	21,000
17	55	F	son	18	25	23	7	2		
18	34	M	Single person	27	27	n/a	1	1	9,250	25,000
19	82	M	2 persons	25	53	25	6	1	850	21,000
20	81	F		24	52	24	6	1		
21	52	F	Widower	20	21	20	3	2	500	23,000
22	38	F	Single person	20	35	n/a	10	1	20,750	23,000
23	42	F	With 2 school age children	23	23	23	2	2	2,995	24,000
24	38	M	With 3 school	23	23	23	3	2	2,500	22,000
25	40	F	age children	25	25	25	2	2		

Orientation to locality	Ever received housing subsidy from employer	Job history	Present occupation	Father's occupation
Nato-local	No	Domestic	Housewife	Army RSM
Nato-local	No	Stable/Low Status	Bank Clerk	SE Builder
		Mixed/Insecure	Taxi Driver	Upholsterer
Nato-local	No	Occasional Supple-mentary Earner	Housewife	Mechanic
Neo-local	No	Services Career	Retired	Teacher
		Domestic	Housewife	Railway Worker
Nato-local	No	Skilled Manual	Retired	Blacksmith
		Domestic	Housewife	Tanner
Migrants	No	Skilled Manual	Technician	Cowman
		Occasional Supple-mentary Earner	P/T Hospital Worker	Farm Worker
Neo-local	No	Semi-skilled Man.	Invalid	Sheet Metal Worker
		Unskilled Manual	Housewife	Rag & Bone Man
Nato-local	No	Unskilled Manual	Retired	Blacksmith
Nato-local	No	Stable/Low Status	Retired	Clerk
		Domestic	Housewife	Mill Worker
Neo-local	No	Deskilled Manual	Chargehand	General Labourer
		Unskilled Manual	Housewife	Farm Labourer
Neo-local	No	Insecure Com.	Estimator	Railway Worker
Nato-local	No	Insecure Com.	Retired	Skilled Craftsman
		Unskilled Manual	Housewife	Factory Worker
Nato-local	No	Unskilled Manual	P/T Factory Work.	General Labourer
Migrant	No	Junior Executive Career	Civil Servant	Policeman
Neo-local	No	Insecure Com.	Audio Typist	Railway Worker
Neo-local	No	Deskilled Manual	Unemployed	Marine Engineer
		Domestic	Housewife	Invalid

even abroad in connection with employment. In contrast very few of those in the working-class area were in career occupations. Few had qualifications or skills which gave them significant bargaining power in the labour market and some had experienced the effects of an interruption in their original employment.

The data in the tables summarize some of the principal characteristics of the housing histories of those interviewed in the two localities. Some of the sharpest contrasts are unambiguously associated with differences in income and occupational status. Others are complicated by the greater diversity of life cycle stages in the working-class area. Nevertheless, while a more general assessment of the changes in that locality would point to some limited infiltration by younger, middle-class households, it remains overwhelmingly an area of residence for a local working class.

Only one couple in the working-class area had always been owner occupiers – and in both cases their parents had been homeowners. This was, however, unusual. More typically, parents had been in rented or tied accommodation and only three of the couples interviewed had bought houses at marriage. The older people in this area had experienced a marked improvement in their housing conditions during their lives. Their most common tenure experience in childhood and as adults (prior to ownership) was of privately rented or tied accommodation. For most of the couples interviewed their current house was the only one they had owned and only two couples had ever owned more than one dwelling. None had ever had second homes or 'investment' properties. The majority of those who had married had lived in two or fewer dwellings since marriage. With the exception of a professional soldier who subsequently entered the civil service and two young single persons in white-collar jobs, those living in the working-class area were in jobs with no (or very limited) career or promotion prospects. Job histories included experience of redundancy and deskilling, periods of insecurity and rapid job change as well as early retirement through ill-health. Employment histories were characterized more by coping than career strategies although the relationship between instability and insecurity varied according to general labour market conditions. Some of the older residents recalled periods in their job histories when bankruptcies and takeovers had thrown them out of work – but this seemed to be part of normal experience with short- rather than long-term unemployment being the expectation and the reality. And contrary to popular and academic belief, stable and secure employment was not necessarily a precondition for house purchase. One of the older households recalled the time they bought their first (and only) house in the 1950s:

> Oh yes, I was out of work when we bought this place – another firm had gone bankrupt. I've been out of work three or four times but for very short periods (authors' transcripts).

They paid £850 outright.

In contrast those in the younger and middle-aged categories, were more likely to see housing and job security closely interlinked and to view unemployment as a structural long-term threat rather than something short term and frictional.

While job histories reflected the discontinuities and volatility of the labour market over a long period, the general picture of the local housing market was of stability and continuity. It remained, on the whole, a close knit working-class community. With only two exceptions, those interviewed were living in their nato locality or were born in an adjacent working-class part of Bristol (neo-locals). (And typically those classified as neo-local households were a mix of neo and nato.) The two migrant households consisted of a couple who had come from a rural part of Ireland and a younger, more mobile junior civil servant – the only person interviewed who was 'passing through'. Property values had risen, but not spectacularly, despite the area being popular and dwellings well maintained. A house purchased for £7/8,000 in the late 1950s was now worth (in 1985) about £22,000.

For those in the affluent area it was relatively common for the parental home to have been owner occupied. There was, however, a considerable experience of tied housing and private renting. None of those in the wealthy area moved into their own home as owner occupiers prior to marriage. And not all became owner occupiers on marriage. An early start in owner occupation was not a common feature of this group. Nor was it the case that these households had been involved in a whole series of moves 'climbing the housing ladder'. For one household their present dwelling was the only one ever owned and for six others only two dwellings had ever been bought. The wealthy area did, however, include households with a long experience of buying and selling and of a significant escalation in the value of the dwellings purchased. Some households had bought a sequence of six dwellings. In each of these cases the principal wage earner was a business executive benefiting from considerable assistance with the costs of moving home. The various forms of assistance on offer through employment emerged as a significant feature of the housing histories of the affluent owners and one of the major factors differentiating them from the working-class histories. The difficulties of attracting senior staff to the high priced South-east had generated a range of complex schemes designed to subsidize the costs of moving. But such schemes were clearly not limited to removal expenses or to moves to London. One executive described the scheme in his company as follows:

> The total amount the company would subsidise you was £25,000. We had a mortgage of £5,000 from the previous two and we therefore had £20,000 to play with. As I said, initially the allowance was for four years and you got a lump sum on the anniversary of moving into your new house. The company then changed that to pay it on a monthly basis on your salary slip. They also increased the period from four years to seven years, but for the fifth and sixth years it is at a reduced rate so we then had an allowance of £5,000 for a four-year period, the fifth year it was reduced by a third and the sixth year it was reduced by a third, so the seventh year it disappeared. That was to ease the pain of losing it. So the total amount you could have was £25,000, therefore we had £20,000 to go. We had our house on the market in . . . [place] . . . for £43,000 but we were actually looking for properties up to £63,000 and we saw various properties. The only one that was within the range was this bungalow that we bought. So we didn't use all the allowance up. Now also the company then changed its policy with regard to expenses and started to give you two allowances, one was 20 per cent of your salary for carpets and curtains, which was non-taxable, there is a tax cut-off point, if you are entitled to £2,000 and you can only have £1,400 tax free, then for the other £600 they would give you £900, as you would lose the tax allowance. In addition to that one, they give another allowance which was 15 per cent of your basic salary which was totally taxable and that was for the intangible costs of moving. Your 20 per cent one was for carpets, curtains and everything else, the other one was for the intangible items like disturbance, moving into a new area and also it would pay any school uniforms as a separate item because we had kids who were then going into different schools and of course, depending on what schools the kids were in the school uniforms could come to quite a bit (authors' transcripts).

While not all households in the affluent area had received such extensive employer-related assistance, most had received some form of relocation subsidy at various points in their housing histories. Indeed, the only exceptions were those in public sector professions or with their own businesses.

Individual histories point up various aspects of labour and housing market links. Two contrasting histories can be referred to as examples. Neither are 'typical' of the two areas concerned but both demonstrate the dependency between housing position and labour market experience. Thus, household W living in the working-class area and household A in the affluent area were at the same stage in the family cycle and were both families with children. In both cases the household included two parents aged in their forties. Unlike many others in the working-class area Mr W had an early job career with an insurance company away from his (and his partner's) Bristol origins. Mr W left home to get married and moved with his job to Middlesex where they rented a bottom floor flat with shared bathroom and toilet. Eighteen months later another job move took them to Hounslow where they rented another flat, this time self-contained. He was offered promotion with the same firm two years later, if they were prepared to move to Luton. He accepted the new post and the company assisted with relocation and removal expenses. It was at this stage that they bought their first house, for £2,250. Her parents provided an interest-free loan to help with their deposit for a small detached property on a new estate. Things, however, did not work out as they had hoped and he left the company four years later. They moved back 'home' to Bristol and reinvested most of the proceeds from their previous house in a three bedroomed newish chalet bungalow in the southern suburbs. They paid £6,750 for it. He found a job as a self-employed agent on the fringes of the finance business. To help out she took a job as a part-time cashier at a further education college. By now they had two children, a son aged two and a daughter aged five.

Working as a self-employed financial agent was both insecure and unrewarding. In 1975 Mr W got a job as a taxi driver and a few years later became self-employed again when he obtained his own Hackney carriage licence. At the time of the interview he had been a taxi driver for ten years. They made their most recent housing move in 1978. This was a move back to the matrilocal area.

> My mum lives three doors away and my brother lives across the road. And I've got an aunt and uncle that live just down the road (authors' transcripts).

Such a move had been partly planned in the sense that her parents were getting older and they had decided to move nearer them. But they had not been seeking a house in the same street. She had mentioned to her father that they were thinking of moving to be nearer the family.

> The next day he phoned me up and said the chap down the road was moving. The gentleman wanted cash in a hurry, our house was almost sold and we were in a position to buy it quickly and at the price he was selling we couldn't refuse it – simple as that. We had been looking slightly further afield really (authors' transcripts).

The house they bought was a three bedroomed Victorian terrace. They paid £11,000. In 1985 it was worth about £22,000. Their move back to familiar territory was not solely to be nearer her ageing parents. They had never really settled in the southern suburbs. They wanted to be nearer the city centre, nearer friends and to have better access to clubs and amenities for themselves and their children, who were approaching secondary school age.

The early housing history of Mr and Mrs A was not so different from that described above and Mr A had entered banking as a trainee on leaving school and had lived with his parents until promotion forced a move to Edinburgh in 1967. There he had lived in digs for a year where he met his partner who was at teacher training college. Their first joint home was a typical Edinburgh Victorian flat with four rooms. They paid £3,600 for it. Two and a half years later they moved to a five roomed semi-detached house built in the 1950s. They paid £6,100 for this house, having sold their first flat for £4,100, £500 more than they had paid for it. They lived there for six years. In 1977 Mr A was promoted to branch manager, a career progression which necessitated a move to London. Property values in Edinburgh (and elsewhere) had risen dramatically in the early 1970s and they sold their house easily for £22,000.

Despite a substantial rise in their household income, the high housing costs in the London market pushed them well out into the Sussex countryside in order to maintain their housing standards. They bought a recently built four bedroomed, detached house near a small village. They paid £35,750, representing a substantial leap in their house price trajectory and in their housing costs. When their next move came in 1981, however, house price inflation in Sussex had increased the value of their property to £87,750. It was at this stage they moved to Bristol so that he could take up an appointment as regional manager.

They expected to make the next career move within four years and static or falling property values had become a major preoccupation. Given his position in the bank the next move was most likely to be London and the head office.

> If I were moved back to that area now, I could have serious problems if I wanted to maintain my standard of housing (authors' transcripts).

But the dominant feature of this housing history was of enhanced rather than inhibited mobility. His rapid promotion in a sector where employees have access to low interest mortgages had produced a steep and highly advantaged ascent of the housing market. All their moves, barring one, had involved payment by the employer of removal costs, legal costs, bridging loans (if necessary), and contributions towards any rental or hotel costs incurred while searching for a new house. Most significantly, however, borrowing had always been cheap enabling them to make substantial leaps in house prices. A move to London would also attract an additional subsidy to compensate for house price differentials.

Apart from some initial similarities, these two housing histories are in marked contrast. In employment terms the Ws could be seen as failed outmovers from the local labour market. After a relatively short foray, seven years, they returned to more hospitable and familiar territory. A possible career in the financial sector led nowhere and from an employee with a larger company with a promotion structure he moved to self-employment and relative insecurity. Their job situation made them reluctant to incur unnecessarily large debts and mortgage commitments had been kept to a minimum on their return to Bristol. Their point of departure in the owner-occupied market had not been far removed from the As. In money terms they started about £1,400 apart. Eighteen years later the valuation gap between them was £90,000. Attachment to locality and being close to friends and relatives would seem to have played a stronger role in shaping the Ws housing history. But if his job in finance had progressed differently they could well have had an equally nomadic history and moved further up the price ladder. The return to Bristol was not a necessary and inevitable move. He could have been a self-employed taxi driver anywhere. Self-employment, however, gave them a choice of location and enabled them to re-establish kinship ties. Moreover, becoming a self-employed taxi driver was more easily achieved in a familiar environment with a supportive social network. As a housing history it would seem to have been a downward trajectory but that was not necessarily their view and it wasn't evaluated in financial terms.

The rate and the geography of residential mobility was one of the striking contrasts between the two groups taken as a whole. Those in the affluent area were more likely than those in the working-class area to have moved house as children, prior to marriage, after marriage before becoming owner occupiers and as owner occupiers. But in this respect the most obvious difference is their greater rate of movement *since* marriage and as owner occupiers. A major element in this difference was job history. In the wealthy area all of the households had embarked upon executive or professional careers or had family businesses. The history of their housing moves was largely a history of job changes which required movement between cities. Indeed, the most striking impression from the interviews was that housing histories were most strongly shaped by job factors and job history. Factors such as parental tenure, age at marriage when children were born, the duration of multiple earnings, early or late entry to home ownership or the number of moves as home owners did not emerge as determining or necessary factors – rather they were component and contingent elements. The necessary factors are principally connected with employment.

In comparing these two groups of home owners there are clear differences associated with such factors as the number and geography of housing moves, parental tenures and continuity of tenures, attitudes towards debt and level of education. There are, however, a number of important similarities which have shaped their individual housing experiences. Among these are the general improvement of housing circumstances over time; the effects of national 'events' such as war; the restructuring of housing tenure; sectoral growth and labour market change and the general dominance of 'use' considerations of shelter and housing over investment or exchange criteria. Even among those in the affluent area who had made a rapid ascent of the housing market there was little evidence to indicate a determined and conscious pursuit of investment and speculation in housing. Affluent migrants were preoccupied with the need to maintain a high level of housing investment because of likely career moves to the London area. Actual decisions to move were most commonly triggered by job factors and from necessity rather than choice. And decisions about *what* to move to and *where* to move to were determined by family and 'use' rather than exchange criteria. It would, however, be wrong to claim that the investment potential of home ownership was unimportant. It is necessary, however, to distinguish the factors which influence specific housing decisions from the primary determinants of residential movement or non-movement. Decisions to move (or not to move) may not be investment driven (see Farmer and Barrell, 1981, for a contrasting view) but financial/investment factors coalesce with other considerations such as house size and location, children's education and social and physical environment to shape housing histories at particular points. Moreover, the financial strategies pursued by individuals in relation to home ownership reflect their labour market position and job histories. One of the owners in the affluent area had never moved voluntarily but when forced to do so for job reasons had made rational economic decisions anticipating the need for further moves (perhaps to higher price areas) and planned early retirement. When asked about the advantages of home ownership, he responded

> I've got to say capital appreciation, I mean that I can't think of any investment that is better. I'd like to be able to talk about peace of mind and all that rubbish, but capital appreciation is what it's all about, which does bring peace of mind. We're going to move down into . . . ninety thousand quid for an end terrace house, two bathrooms and so on. But as a result of having done what we've done I've got another £50,000 to eke out our pension (authors' transcripts).

The point is that the nature of the housing market had delivered significant benefits to this household but the overall shape of their housing history would have had the same features (in terms of geography, number of moves, etc.) regardless of the tenure structure. In other words, they moved *through*, not *because of*, the owner-occupied market. And their attitudes towards housing and their specific housing market trajectory are made sense of in relation to employment. This relationship is as strong among those in the working-class area where limited residential mobility and low levels of housing investment are also associated with aspects of employment. Financial strategies in this context tend to revolve around minimizing indebtedness and escaping from private landlordism. Peace of mind and security come from outright ownership and the knowledge that disruption of employment will not threaten housing circumstances. An elderly husband in the area had been a traditional grocer in his early career and as that trade disappeared had become a sales representative. He recalled a volatile employment history and a long series of similar jobs in various companies (as well as the fact he had the only car in the street when he started in sales). For him the advantages of home ownership were that:

> Nobody can take it away from you or kick you out and it's a nice feeling once you've paid for it. You're not owing to nobody. I only took it [a mortgage] out for fifteen years, which was a struggle at first, but as the years went on and I got more money I got embarrassed taking my account in. I was only paying £9 a month and people starting were paying

£20 or £40. They must have thought 'what's he living in then?' (authors' transcripts).

One very obvious feature which emerges from carrying out this kind of research is that snapshot surveys of attitudes towards tenure can be quite misleading. For example, many of the responses to the question 'What do you see as the advantages of home ownership?' would fall into the 'financial' category and be grouped together in a conventional report of a social survey. But different households have quite different financial orientations and strategies. Attitudes towards, and the meanings attached to, housing tenures can only be made sense of in the light of past experiences and cannot be assumed to reflect some commonly held values and beliefs. Few people have coherent images of the range of tenures on offer. People offer views of tenure reflecting their own experience as adults and children. For some this is exclusively an experience of home ownership. For others the reference point is tied accommodation or private renting. Very few 'affluent' owners had had experience of council housing. And for some renting was associated with higher costs or tied housing with very regular and disruptive movement. What was most evident, however, was the enormous gulf between the housing histories of those whose housing experiences had been shaped by the distinctive features of careers associated with business executives, salaried professionals, the professional armed forces and family businesses in contrast to those whose skills rendered them easily substitutable as homogenized labour and who had relatively weak bargaining power in the labour market. This is not to suggest, however, a simple division between 'successful' and 'unsuccessful' housing and job histories or a class divide of advantage or disadvantage. There *was* a strong class dimension and clear differences in terms of income and wealth but it is in consideration of the finer detail where some of the more interesting differentiating features emerge and which undermine a view of a unitary value system or a simple relationship between home and work with success or failure judged against a common set of aspirations. Some of these aspects are discussed further in the concluding section.

Housing markets and labour markets: a discussion

From what has been said earlier we are in a position to draw out a number of themes and issues relevant to a consideration of the connections between job and housing histories. It should be emphasized that the study itself did not set out with the primary aim of investigating those connections. Rather it was concerned with the nature of stratification within home ownership and the ways in which households and individuals confront and negotiate complex social structures. The intention, however, was not to generalize about strata within home ownership but to indicate the key factors which appear to generate divergent experiences within two segments at opposite ends of the owner-occupied sector. Moreover, concentrating on households in the same tenure position and on the shaping of their *overall* housing histories avoids the danger of being mesmerized by tenure divisions as the *primary* differentiating features.

Certainly the most striking feature of the housing histories of the groups discussed in this chapter is the ways in which they have been structured by employment imperatives. There is little to suggest that individuals have achieved social mobility through conscious manipulation of the housing market. Access to high status housing and of significant accumulation of housing wealth is restricted either to those with independent wealth or income from family business or self-employment or those in professional or salaried business careers in which career progress is dependent on residential mobility. But in this the relationship between job and housing is not a simple undimensional relationship between income and access to properties of a different price. Job-related housing moves and company assistance with the costs of moving and in some cases the increased costs associated with a new dwelling are a component of certain types of job careers. For those individuals and households, career progress is accompanied by progress in terms of the quality, quantity and value of housing consumed and owned. Jobs with equal status and earnings are differentiated according to the changes in housing circumstances which are a necessary accompaniment of job changes. The type of job rather than simply level of earnings, security or other factors determined the extent of residential mobility.

Occupational housing benefits emerge as a major feature of some housing histories. The value and nature of these benefits through housing and job history differ dramatically. Most obviously, those without professional or other acquired skills, where labour was easily substitutable and who experienced deskilling, had job histories which signally lacked occupational housing assistance. For others, promotion and progress in job terms automatically involved moving house and job assistance with moves. And in some cases, a relatively insecure and meandering employment history revolved around residential immobility and a housing history sedimented in a strong social and kin network. In this sense non-movement as much as movement is indicative of a particular kind of employment history. Housing position and housing histories form part of broader 'ways of life' associated with particular occupational categories and the connections between housing and job histories are more complex than implied by inference from income.

Different occupational groups occupy different parts of the housing market. Those in the affluent area moved in a national housing market constituted by subsections of various local housing markets. Those subsectors, often associated with new executive developments, serve those 'on the move', and are fuelled by a set of financial arrangements and lifestyle criteria which may set them apart from the general contours of the local housing market. In contrast, other areas are essentially local, serving local people and reflecting local economic factors. This distinction is not, however, a simple reflection of class or income. To some extent, there is a distinction between the nomadic 'cosmopolitans' and the local 'burgesses' identified in earlier sociological literature. It is not a simple division between middle-class mobility and the stable and residentially immobile local working classes. For those on the move, life cycle factors connected with locating in particular school catchment areas may be of prime importance. And ease of resale figures much more prominently in the residential decisions of the mobile professionals and executives as opposed to the locally housed affluent who are better able to maximize housing use values. For senior employees with a past and prospective career in a national or international internal labour market, there is a dominant preoccupation with the next move and the need to anticipate the housing consequences of an eventual relocation to a headquarters which is likely to be based in London or elsewhere in the South-east. Some of these dilemmas and considerations are captured in this extract from an interview with a senior executive in a multi-national company:

> . . . if you pick an estate and you pick one with four bed-roomed houses, you assume, rightly or wrongly, that the families that occupy them are going to have children. One of the other things that I have always been conscious of is that in being promoted back into the South-east, you can significantly alter your quality of life. To give you an example, if we had put the boys into a school in Tiverton, we could have gone and lived six miles outside Tiverton and bought a house that was on the market at that time for about £93,000 which was everybody's ideal . . . most people's dream house, it was a sixteenth-century house that had six bedrooms and about four or five rooms besides, in about two acres of land. Now if we had bought that house we would never have left it. And therefore in picking the most expensive area, you don't get the same amount for your money (but) it is relatively easy to

> move from a house like this. Even if it decreased in value to £108,000 it would be much easier to move from this house, like something in the South-east than it would have been from a house in the Wellington area, where for ninety or so thousand pounds you get a mansion with shooting and fishing rights (authors' transcripts).

What all this conveys is the disconnected nature of the housing market. There is not one market with different groups pursuing similar goals and strategies. There is not a unitary value system, a single housing ladder, with the ultimate position being dependent simply on financial bargaining power. Different forms of employment shape housing histories in qualitatively different ways. In some cases, typically higher status employment necessitating more frequent and more geographically diffuse moves, kinship ties are broken and social networks are ephemeral. In others, attachment to kin and/or locality are the dominant influences. Some housing histories span national space, others are highly restricted to small localities. In some cases, the housing history is a history of family change and adjustment, of inmigration and outmigration, of labour market volatility restricted to one residential location. But even where influences other than employment appear to be dominant in housing decisions and experiences it may be the sphere of work which provides the key parameters in which those influences operate. While it is undoubtedly the case that some individuals are more likely to move than others, much of the literature on residential and employment mobility conveys the impression that people in general have a strong and unrelenting desire to move. Patterns are often interpreted as revealed preferences. But among the senior and highly mobile executives and professionals in this study there was little evidence of a desire to move house independent of job constraints. Residential movement was fundamentally a product of 'career path migration' (Salt, 1985). And that career path migration had clear social costs - housing histories shaped and dominated by the imperatives of the male's employment and narrow, instrumental and short-lived social networks. When asked about close friends one male executive responded:

> I don't have any . . . that's one of the problems of moving around such a lot. They're probably in the company you work with. I've really not had a friend since I was totally independent . . . I have acquaintances and associates. That's one of the disadvantages of this sort of career (authors' transcripts).

And another male in a similar senior executive position, observed:

> We are very different from them. You will appreciate we come from a working-class background and we are unique . . . None of the others [relatives] would have moved, to move ten miles would have been a colossal adventure whereas we've moved so many times it isn't true. I was brought up to be extremely close, it's just that we moved away. It just changes things (authors' transcripts).

In contrast, the Mr and Mrs W referred to previously in the working-class area had moved to pursue an early career path for the husband in the financial sector but, after several years away, they returned to their old locality, to a more hospitable and familiar territory. An important dimension of their housing history was that an early series of moves was influenced by *his* employment prospects. On their return to Bristol it was *her* kinship network which influenced their residential location. This was a common feature of housing histories in the working-class locality and contrasted with the male-influenced moves in the affluent area. In this sense, the nature of employment (and not only the gender division of formal paid employment) has an important influence on gender relations and highlights a dimension of housing histories which tends to be underemphasized in the literature which stresses life cycle, family adjustment and relative mobility rates. And where job factors are stated as the reason for movement it is not always clear *whose* job necessitated the move.

Given the qualitative and highly localized nature of the research reported in this chapter some caution is clearly required in interpretation and generalization. Nevertheless, it is our contention that the reconstruction of households' housing histories is a useful and necessary complement to other methods of investigating housing issues and is of particular relevance in exposing more subtle forms of interconnectedness – in this context between labour and housing markets. There is more room for the unexpected, the element of surprise although more systematic and exhaustive research using other methods will be required in order to quantify.

The ways in which employment may affect housing debt strategies, the gender dimension of the shaping of housing histories and the emergence of new forms of occupationally related housing subsidies are examples of issues which are rarely discussed in the mainstream housing literature. More substantial evidence on relocation assistance is now emerging (see, for example, IMS 1987). Moreover, this research suggests that explanations for differing migration and mobility patterns, and for the evident divisions in the housing market, are to be found in processes of labour market segmentation, the operation of internal labour markets and the widening of the social divisions between those in employment and those on the social and economic margins. Rather than starting with housing tenure divisions it may be more appropriate to conceptualize distinctions in housing opportunity and experience in terms of a housing market segmentation which relates to segmentation in the labour market. There may be a housing market division which corresponds to the familiar distinction between primary and secondary labour markets. Although such a bifurcation of the labour market is generally recognized as requiring considerable qualification and refinement it, nevertheless, may represent a useful point of analytical departure. Past research established strong connections between employment in the secondary labour market, low geographic mobility and the public housing sector. Recession, economic restructuring and tenure restructuring (in particular the progressive penetration of home ownership down the income structure) indicate the need to reappraise those relationships.

As the unemployed and economically inactive are increasingly concentrated in the rental tenures, labour market divisions may now be of more relevance in explaining contrasting trajectories *within* home ownership. Historically, the clearest connection between housing and employment was in the provision of company dwellings and tied accommodation. The re-emergence of interest in housing and labour market links should take into account how company personnel policies reflect and act upon the dynamics of the owner-occupied market. Company recruitment and retention policies may become increasingly significant in the shaping of housing histories as internal labour markets continue to expand with the concentration and centralization of capital. And if market orientation in housing provision continues to increase, those on the periphery and those excluded from employment are likely to be excluded also from housing advantages directly linked to employment. An understanding of future and emerging patterns of residential and urban differentiation will require more sensitive and detailed analysis of the ways in which different groups in different labour market positions, in different jobs and employed by different companies, have a different experience of the housing market.

References

Bateman, M. and Burtenshaw, D. (1973), 'Sponsored Migration and Residential Change'. Paper presented to IBG Studies Group Meeting, Sheffield.

Bertaux, D. (ed) (1981), *Biography and Society*, (London, Sage).

CBI Employee Relocation Council (1986), *Relocation News No. 1*.

Davis, N. (1986), 'Training for change' in P. E. Hart (ed.), *Unemployment and Labour Market Policies* (London, Gower).
Dunleavy, P. (1980), *Urban Political Analysis* (London, Macmillan).

Edel, M. (1982), 'Home ownership and working class unity', *International Journal of Urban and Regional Research*, 6, 4, pp. 481–502.

Farmer, M. and Barrell, R. (1981), 'Entrepreneurship and government policy', *Journal of Public Policy*, vol. 2, pp. 307–32.
Field, F. (1981), *Inequality in Britain: Freedom, Welfare and the State* (London, Fontana).
Forrest, R. (1984), 'The Meaning of Home Ownership', *Society and Space*, 1, pp. 205–16.
Forrest, R. and Murie, A. (1987), 'The affluent owner: labour market position and the shaping of housing histories', *Sociological Review*, 35, 2, pp. 370–403.

Gray, F. (1982), 'Owner-occupation and social relations' in S. Merrett with F. Gray, *Owner Occupation in Britain* (London, Routledge and Kegan Paul).
Green, F., Hadjimatheou, G., Smail, R. (1984), 'Unequal fringes', *Occasional Papers in Social Administration*, 75 (London, Bedford Square Press).

Harloe, M. (1984), 'Sector and class: a critical comment', *International Journal of Urban and Regional Research*, pp. 228–37.
Holmans, A. (1981), 'The housing careers of recently married couples', *Population Trends*.
Holme, A. (1985), *Housing and Young Families in East London* (London, Routledge and Kegan Paul).

Ineichen, B. (1981), 'The housing decisions of young people', *British Journal of Sociology*, vol. 32, no. 2.
Institute of Management Studies (1987), *Relocating Managers and Professional Staff*, IMS Report no. 139, Institute of Management Studies.

Merrill Lynch Relation Management International Limited (1986), *Third Annual Study of Employee Relation Policies Among Major UK Companies*.
Mitchell, J. Clyde (1983), 'Case and situation analysis', *Sociological Review*, 31, 2, pp. 187–207.

Parsons, D. (1987), 'Recruitment difficulties and the housing market', *The Planner*, vol. 73, pp. 30–4.
Plummer, K. (1983), *Documents of Life* (London, Allen and Unwin).

Pratt, G. (1982), 'Class analysis and urban domestic property: a critical re-examination', *International Journal of Urban and Regional Research*, 6, 4, pp. 481–502.

Salt, J. (1985), 'Housing and Labour Migration'. Paper presented at the Conference on Housing and Labour Market Change, Parsifal College, London, December.

Saunders, P. (1984), 'Beyond housing classes', *International Journal of Urban and Regional Research*, 8, pp. 202–27.

Thorns, D. (1985), 'Age time and calendar time', *Environment and Planning A*, 17, 6, pp. 829–44.

3 *Labour migration and housing in the UK: an overview*

JOHN SALT

Introduction

Contemporary attitudes to labour migration in the UK and other developed economies are confusing. Quite often those who have a 'need' to move – predominantly the unemployed – are unwilling or unable to do so. Conversely, those whose positions in the labour market are strong, being in highly skilled and well-paid occupations, may move often; indeed, for many people geographical mobility is a way of life. The attitude of the British government towards labour migration is anything but clear. Ministerial exhortations to migrate are accompanied by actions which tend to curb movement. The present government view is still largely epitomized by the concept of 'bikeism', introduced by Norman Tebbit in 1981 in a much misquoted and misinterpreted speech to the Conservative Party Conference.

> We cannot ignore the price that unemployment today is exacting. I know those problems. I grew up in the Thirties with an unemployed father. He didn't riot. He got on his bike and looked for work and he kept looking until he found it (Tebbit, 1981).

Mr Tebbit was echoing an earlier view expressed by the Prime Minister of the need to reduce barriers to geographical mobility.

> Frequently investment goes where there are skilled people wanting work. But there must be some mobility of labour. If people are not willing to move as their fathers did the economy cannot thrive (Thatcher, 1980).

Despite this verbal exhortation early in its first term of office, the present government has since said and done little to encourage labour migration, adopting a *laissez faire* attitude. Measures to facilitate labour migration do not feature in the Manpower Service Commission's programme and the 1986 budget abolished the only government mobility of labour assistance scheme – the Employment Transfer Scheme – which had existed in one form or another since 1927. Currently the responsible UK government departments are engaged in no major studies of labour migration and, so far as one can tell, there is no clear policy aim as far as long distance movement of the British population is concerned.

Official reticence is matched by relative unconcern in the limited academic literature on the subject of labour migration. The neo-classical equilibrium view still largely holds and the amount of empirical research in recent years has been small. There has, for example, been no major survey of labour migrants in the UK since the 'Housing and Labour Mobility Study' in 1972 (Johnson, Salt and Wood, 1974). The growing literature on regional unemployment and deindustrialization largely ignores the role of labour migration. One of the most highly regarded books on the new spatial division of labour (Massey, 1984) scarcely mentions labour migration, even in passing.

One of the few accepted 'truths' about labour migration is that it is inhibited by 'housing' as testified by a growing body of literature (for example, Hamnett, 1984; Conway and Ramsey, 1986; Minford, Peel and Ashton, 1987). What is not clear, however, is exactly how and to what degree housing difficulties constitute a barrier to geographical mobility. Tenurial problems, especially affecting renters, and regional price differences, affecting owner occupiers, undoubtedly exercise an arthritic effect. The principal housing barriers to mobility are discussed later. Yet housing remains merely one among several barriers confronting the potential mover, and their ability to surmount each of these varies. Most obviously, jobs in potential destinations need to be available in branches of work for which the potential migrant has the requisite qualifications. At times of high unemployment – like now – the likelihood of this being the case is considerably diminished. This means that, despite the existence of marked regional variations in employment, for the out of work especially, but also for the person in work seeking a new employer elsewhere, the likely gains from movement are limited. The potential migrant also needs *information* on available opportunities and has to cope with the steep distance decay effect in this regard. This constraint is less for those whose jobs tend to be advertised nationally, in newspapers or specialist journals, and who ultimately move to a job and destination fixed up in advance. The *costs* of moving constitute another major problem.

These may be the simple physical costs – fares and removals. More important for owner occupiers are the costs associated with housing change – buying, selling and equipping. House price gradients and stagnant markets, necessitating expensive bridging loans, make the situation worse. For those renting homes there are often difficulties of availability and qualification. There are also psychological costs, often associated with the others, which may end up in divorce, illness or even premature death.

A fourth set of problems is associated with the range of personal and environmental *ties* that migration will break – family, friends, schools, a favoured area. Finally, there are problems of assessing *future prospects*, especially those of career and yet further movement. Only if prospects seem reasonably assured will the migrant household deem the trauma of moving worthwhile.

The thesis of this chapter is that the labour market creates conditions in which labour migration is deemed desirable from both the supply and demand side. It is a process which is occupationally selective and takes place in a policy framework set by public (government) and private (employer) organizations. Housing presents one set of obstacles, the nature of which is changing. The response has been a growing polarization between those who can more easily surmount barriers and whose occupations make movement easier and others for whom the reverse applies. Although the perspective is very much a British one many of the issues raised have wider relevance in advanced economies; this applies especially to the theoretical ideas developed. The chapter proceeds by defining labour migration and then reviewing the significance of the labour migration process for population and employment. A discussion of labour migration theory precedes a comment on contemporary patterns of migration within the UK and the policy framework in which they occur. The next section looks at an increasingly significant element in labour migration within developed countries, that within internal labour markets (ILMs) of large employing organizations. Finally the nature of housing difficulties is discussed.

Labour migration: a definition

There is no commonly agreed definition of what constitutes labour migration. A frequent assumption seems to be that those involved move primarily for employment reasons, usually as stated in questionnaire surveys; alternatively definitions are sometimes based on distance moved, with interregional migration being regarded as a surrogate. Several writers have sought to distinguish between labour- and housing-inspired migration (Bonnar, 1979; Gordon,

1979; Gordon and Vickerman, 1979), the latter being essentially over short distances. Definitions based on distance make assumptions about commuting tolerance and the geographical scale of the local labour market. Bonnar (1979), for example, has argued that long-distance housing adjustment in South-east England makes it difficult to differentiate between the two. The situation is complicated further by weekly commuting from depressed parts of the country into southern England; a practice which has grown considerably in recent years (Policy Studies Institute, 1987). A labour migrant is probably best defined as someone who simultaneously moves both home and job location between local labour markets (Johnson, Salt and Wood, 1974), and this definition is used here. This coincidence occurs when the relocation of either job or home leads to the overstretching of commuting tolerance, necessitating a change in the other. Such a definition does not rely on motivation and a labour migrant may thus be someone who moves primarily for housing, personal or other reasons as well as employment. This is logical because any migrant who is in the active labour force age range is either a worker or potential worker. His or her migration has an actual or potential impact on the labour market and should be regarded as a labour migration.

Regional significance of labour migration for population and employment

Population

The period of relatively rapid natural increase in population in the UK since the end of World War II has ended, and immigration from overseas has been less than emigration until the last few years. As a result, internal migration of all types is even more important as the main cause of change in local population totals. The impact of internal migration is not confined to total population, for regional demographic structures are also changed. Most inter-urban movement consists of exchanges of population which may be qualitatively selective, bringing about dilution of skills in some areas while concentrating them elsewhere. Furthermore, the rate of turnover is important, having implications for the vacancy rates in housing and labour markets, for example. Selectivity in migration is also creating new land-use planning problems and accentuating old ones. The pressures on land and housing in the corridor of growth to the west of London are calling into question Green Belt policies and the slowing down of out-migration from major metropolitan centres (and its reversal in recent US data) suggests new attitudes may be needed to inner city problems. Finally, past and prospective trends in population redistribution raise major issues of the sort of urban system that is best for Britain and of the policies that are appropriate to achieve it.

Employment

As the UK tentatively emerges from the economic recession of the early 1980s, it is becoming clear that economic growth does not necessarily lead to increasing employment. Even if growth is sustained with any consistency, new jobs in manufacturing may prove to be much less footloose in their location at the regional level than in the past, since, with the new international spatial division of labour in which multinational firms play a major role, the marginal benefits that can be derived in the economically more attractive parts of Britain cannot be ignored. Employment in the service sector as a whole, but especially in private services, can also be expected to change its location and composition. While jobs in producer services have increased, new technology may also bring about a decline in this sector in the long run. Certainly, current technological development seems likely to continue the decentralization of this form of employment though new forms of communication (such as fibre optic cable 'motorways') may bring about change in new directions. The greater significance of non-wage benefits of labour, such as high skill levels and low unionization, are also likely to influence relations between employment and labour mobility, so that traditional associations of production and labour costs become less important. In sum, it seems unlikely that the location of jobs lost as a result of more capital-intensive production and the disappearance of declining industries will match that of the new jobs being created in service employment and in technologically advanced manufacturing. Hence, labour migration and occupational mobility will be essential if structural mismatch is to be reduced and if the full potential of Britain's resources is to be realized.

Contemporary labour migration in Britain

Between 1971 and 1985 National Health Service registration data indicate that migration both between and within regions declined by an average of between 2 and 2.5 per cent per annum, a figure consistent with census data which show an average annual decline of 2.4 per cent in both inter- and intra-regional movement during 1971–81. Nevertheless, with 922,000 inter-regional moves in the

UK in 1985–6, according to the NHS register, a lot of people do manage to migrate.

Most people, of course, move short distances. For 1980–1 the census records that only a quarter of all movers went beyond the immediate local area (greater than 20 kilometres), though over half of these migrated over 80 km. This higher propensity to move is reflected in mobility rates; over 20 kilometres and over 80 kilometres professional and managerial workers had higher mobility rates than junior and intermediate non-manual and manual workers (33.1 per cent of professional and managerial migrants go over 20 kilometres; 28.25 per cent over 80 kilometres; equivalent figures for junior and intermediate non-manual workers were 23.7 per cent and 12.7 per cent, and for manual workers 11.1 per cent and 6 per cent). Propensity to move over long distances also varies by industry. Greatest mobility is in the 'Other Services' group recorded in the census, which includes national and local government, education and health; sectors which are predominantly white collar and have a high built-in mobility related to career development and widespread branch networks. A similar situation applies in insurance, banking and finance. In contrast manufacturing is associated with low long-distance mobility rates, as most of the workforce there consists of manual workers in occupations without built-in career systems and with little opportunity or incentive to move. Similar conditions apply in the low mobility transport, extraction, and energy and water sectors. Construction might be expected to involve considerable mobility given the 'Navvy' image, but most construction workers are in small firms doing business locally and with little reason to move far.

What has become apparent only recently is that a substantial proportion of labour migrants are transferred by their employers. The 1981 Labour Force Survey records inter-regional movers aged over sixteen during the preceding year, and whether or not they were working for the same employer before and after the move. Of 582,113 inter-regional movers during 1980–1, 248,273 (42.7 per cent) were employed at both dates. Of this second group 144,930 (58.4 per cent) were employed by the same organization before and after moving. The author's own Organizational Labour Migration Survey data (see p. 103) suggests that about 180,000 labour migrants per annum in the country as a whole are transferees; with an average household size of 2.2 people this means that about 400,000 people annually move under company auspices. Atkinson (1987) has recently suggested that the number of transferees (excluding family members) is now 250,000 per annum.

Theoretical explanations of socio-economic migration differentials

Socio-economic differences in migration propensity occur in all developed economies. These differences reflect the segmentation of the labour market between socio-economic groups, and hence explanations of these differences require consideration of labour force heterogeneity. The classical economic approach to inter-regional migration (Lind, 1969) stresses its role in the adjustment of supply and demand for labour. If supply exceeds demand in one region, wages there will fall or unemployment will rise; where demand exceeds supply, full employment will ensue and wages will rise. One effect of this may be to encourage entrepreneurs to locate in the former region, where labour is available and cheap. Another effect is to induce migration; unemployed workers can improve their chances of employment by migration to a region of labour demand, and employed workers can benefit from migration by moving to a region of high wages. If there is universal high unemployment, as recently, inter-regional labour migration can have only a marginal effect (Green, *et al.*, 1986).

This argument can be extended to take account of occupational specialization. Migration is induced by a regional mismatch of supply and demand in a particular occupation. If each occupation has a totally separate labour market, independent migration streams will be generated to produce separate regional equilibria for each occupation. It would therefore be quite possible for a large stream of migrants in one occupation to move from one region to another while migrants in a different occupation moved in the reverse direction. In practice, occupational labour markets are not totally discrete, and workers in one region may be able to find jobs or better wages by changing occupation rather than by inter-regional migration; Gleave and Palmer (1980) have shown how occupational and geographical mobility vary in importance between occupations.

The implications of these arguments are that migration should in general be from regions of low wages and high unemployment to regions of high wages and low unemployment, and that it should be highest in those occupations where regional differentials are greatest. The latter are likely to be those occupations whose regional pattern is becoming more concentrated or more dispersed. Ease of entry to an occupation may also affect migration patterns; if it is hard for workers to move into a particular occupation, normally because of training or qualifications deemed necessary, regional shortages must be filled primarily through immigration, while vacancies in an occupation requiring little training can be filled through occupational rather than geographical mobility (Johnson, Salt and Wood, 1975). Generally, therefore, inter-regional migration rates might be expected to be lower for unskilled than for skilled or professional workers, a conclusion supported for the UK by analysis of occupational migration rates (Salt and Flowerdew, 1986).

A further economic factor related to occupational migration is the cost of movement, now estimated at £7,000–£10,000 for an owner occupier (Salt, 1986). Migration is costly, not only because of the transport costs involved, but also because of the transaction costs, especially where a house must be sold and another bought. Regardless of tenure, there may be a need for a migrant to maintain two houses for a period or to use temporary accommodation at a high price. The expenses of migration may vary according to distance and according to the migrant's life cycle stage, a single person finding it far easier to migrate than a member of a large household. It is also the case that employers are more likely to assist some migrants than others with their expenses; this assistance is in general higher for occupations where more training is required, although such people are more likely to be able to bear the costs themselves (Salt, 1984).

There are also sociological arguments which suggest that occupational differences in migration rates might be expected. These are concerned with the relative importance of employment and other factors, especially attachment to family and community, and how it varies between classes and occupations. Migration will break a wide range of personal and environmental ties, perhaps involving family, friends, schools and social activities. The importance of these ties may depend on previous migration experience and on the perceived ability to maintain them despite physical distance, both of which are likely to be related to social class and hence to occupation (see, for example, Jansen, 1969; Taylor, 1969; Gleave and Palmer, 1980).

In addition to traditional economic and sociological explanations for migration a range of behavioural theories has been developed emphasizing the nature of search processes (for alternative localities, jobs and homes) and decision making (for example, Wolpert, 1965; Mangalem and Schwarzweller, 1968; Flowerdew, 1976). The assumptions made in behavioural models are more realistic in many respects than in most economic models but they are still variable with respect to information available, whether or not a new job is involved and how many economically active people there are in the household.

In general, behavioural models have assumed that the migrant is choosing a new job and that the decision is made by the individual or migrant household. Largely left out of account is the employer who makes decisions about skill and experience requirements, recruitment procedures and so on. Indeed, perhaps the fundamental weakness of research on labour migration is to study it as a supply-side phenomenon, from the point of view of the migrant. Quite clearly the role of the employer in filling vacancies needs to be incorporated into any theory of labour migration, but how is this to be done?

Labour migration and its attendant occupational selectivity should be viewed in the light of the specific tasks to be performed in any occupation and the role of the occupation in the employer's allocation of tasks. Furthermore, occupations must be viewed in career terms, where 'career' is defined as the incremental accumulation of skill and experience as the individual performs the range of tasks and occupations allocated within the employer's internal labour market. Thus we need to study not only those characteristics of occupations that make them more or less mobile but also how the organization of work affects geographical mobility.

From this perspective, the concepts of internal labour market (ILM) and organizational career become central. An individual in employment is, by definition, in an ILM, working according to its rules and corporate practices, his or her career guided or determined by how the employer chooses to organize the tasks needing to be performed. Anyone recruited externally, moving from one employer to another, is really moving between ILMs, through 'ports of entry' defined in terms of ILM organization. Ports of entry may be at any level in the company hierarchy but their degree of openness will usually be established in accordance with perceived recruitment needs and consequent vacancy-filling procedures and with the company's philosophy on training and staff replacement. The concept of organizational career is also relevant to the phenomenon of immobility. Short career ladders in some occupations provide few opportunities for advancement and hence few occasions when migration can bring career benefits; the labour markets in such occupations are usually spatially restricted.

Several studies have emphasized the importance of geographical transfers of workers within an organization's ILM (McKay and Whitelaw, 1977; Johnson and Salt, 1980; Sell, 1983). Traditional explanations for labour migration are inadequate for explaining this phenomenon and this emphasizes the need for a framework linking migration and the organization of work. The validity of this statement can be seen in the empirical evidence below.

Organizational labour migration

The significance of transfers between the sites of multilocational companies in the US has been documented by Sell (1983) for the 1970s. Using Annual Housing Survey data for 1973–7 he calculated about 800,000 relocations per annum. The only data on the phenomenon available for the UK are from the author's survey of the transfer policies and practices of large multilocational organizations. The study, carried out in the early 1980s, was funded by the ESRC and coincided with the worst of the recent recession (Salt, 1984).

The study investigated ninety-three major organizations, or major parts thereof, and obtained interview data on internal labour market operation and mobility policy, as well as quantitative data on number and characteristics of movers and their geographical pattern. The sample, consisting of multisite companies, mostly with over 5,000 employees in the UK, included organizations from all sectors of the economy: government departments and nationalized industries as well as privately owned organizations. Interviews were carried out between January 1982 and Easter 1983. Numerical data on scale and patterns of movement were not available for all organizations interviewed, nor was the detail provided identical.

Rates of transfer

Internal transfer numbers were available for seventy-six organizations, giving an aggregate annual total of 7,404 moves out of a total labour force of 980,084, and an average of ninety-seven moves per organization per annum. Not surprisingly, there were wide variations around this figure: 35 organizations had fewer than 25 moves per annum (13 less than 20), while 18 had over 100, including 11 with over 200, 3 over 500.

In most cases transfers were in managerial, professional or senior technical grades. Only rarely do companies transfer hourly paid staff; this has always been so. A similar situation also applies to clerical and lower paid technical staff; the main exceptions are married women and young females transferring at their own request to be with mobile husbands and parents. Such movement is most common in those organizations which have extensive branch structures: hence in the case of one bank studied the majority of its migrants were lower grade female clerical staff.

Clearly the number of moves will reflect overall organization size so a Total Mobility Rate (TMR) was calculated, expressed as numbers of transfers per annum as a proportion of total employment. Experience ranged from 3.0 to 0.02, with an average of 0.74

and a median of 0.4. When grouped by broad economic sector, government departments were most mobile (average = 1.67), then finance (average = 1.30), retail (average = 1.01), transport and communications (average = 1.00), manufacturing (average = 0.56), public utilities (average = 0.46), and extractive industries – which include some primary processing manufacturers (average = 0.42). Within each of these categories there were major variations reflecting policy and structural variables, as well as economic conditions and the geographical distribution of locations.

A mobility rate based on total employees can give a misleading impression when only a part of the labour force, normally from the first or second supervisory level upwards, may be considered as at risk of transfer. Accordingly, organizations were asked at interview which of their employees might be regarded as potentially mobile for career purposes (excluding special circumstances such as redundancies). Proportions varied from 64.2 to 1.1, with a mean of 19.9, a median of 15.9, and a modal range of 5.0 to 9.9, the last reflecting the inclusion in the sample of a number of organizations with very high proportions of manual industrial and clerical staff.

Recalculation of mobility rates based only on potentially mobile staff presents a different order from total mobility rate, providing a truer reflection of the mobility ethos of the organization. Eight organizations had mobility rates of 10 or more, the highest being 23.7 (meaning that its potentially mobile staff have more than a 1 in 5 chance of being moved per annum). Other organizations continued to have low mobility levels, the lowest being another retail organization with 0.3. In general retail organizations were most mobile (average = 8.0), then finance (average = 4.2), government departments (average = 4.1), manufacturing (average = 3.8), transport and communications (average = 2.9) and public utilities (average = 2.4). Again there were variations within sectors reflecting policy, structural, geographical and general economic variables.

In some cases organizations had transferred *groups* of people *en masse*, consequent upon the opening or, more usually, closure or relocation of a whole unit of production or an office. Twenty-four such cases were identified and distinguished from those transfers of *individuals* within the organization which might have resulted from rationalization and redundancy. The proportions of moves of this type varied from the very high figure of 98 per cent down to 3 per cent of all transfers.

The organizational career

The analysis thus far has been based on cross-section data relating to the early 1980s. It is important, however, to see the organizational transfer not as an isolated move but as part of the sequence of moves that goes to make up the organizational career.

The concept of organizational career is crucial to explaining migration. From the management viewpoint the internal structure of the organization provides an orderly hierarchy of responsibility and authority, with vertical and horizontal movement of personnel through it. From the employee's viewpoint, the structure provides a stable set of expectations. Over time, paths of personnel tend to become more or less stabilized through the firm as vertical and horizontal movement paths evolve to form career lines. Within the total complex of positions certain ones operate as critical junctures or testing points. The organizational career is thus a relatively standardized set of roles to be performed by an employee which interact with a relatively patterned flow of individuals through these roles (Milkovich, Anderson and Greenhalgh, 1976). Organizational careers have a patterned sequence of positions related to work content and an ordered movement of individuals among these positions.

We may hypothesize that in different organizational career lines propensity to migrate will increase as critical junctures are approached, but when and how frequently do these occur? In addition to the cross-section data, employers were asked for a sample (normally 50–100, depending on size) of career histories within the organization of potentially mobile employees, those whose careers might at some stage carry the risk of movement. Most complied and ultimately over 3,000 career histories were obtained. From these it has been possible to compile, for each individual, every job change in the ILM, occupational and geographical. An occupational move could be to a related or different type of job. Since home address was not available for each job, relocation was based entirely on work location. Moves were subdivided into those over and under 32 kilometres. Geographical mobility, expressed in terms of moves over 32 kilometres as a percentage of all job changes proved to have a surprisingly large role in ILM operation. In the case of the 'most mobile' employer 57.9 per cent of all job changes recorded in the sample of 50 career histories involved a transfer of over 32 kilometres. Even in the most immobile company, a heavy engineering organization for example, getting on for 1 change in 10 was associated with a migration.

These figures raise many fascinating questions about the relationship of mobility frequency with promotion, function, age, seniority and others, which cannot be dealt with here. The salient point, though, is clearly that geographical migration plays a major but varying role in the internal switching of jobs among potentially mobile staff within large organizations. Since mobility of staff is clearly in the operational interests of the employer, housing must not be allowed to become an insurmountable barrier.

Organizational moves and housing

Virtually all of those moving in company transfers are managerial and technical staff and for them generous relocation packages exist which can provide a magic carpet across the bumps and hollows of the housing market. One survey of 243 organizations in 1984 found that 97 per cent compensated transferees for removal expenses; 96 per cent for temporary living and travel costs, 93 per cent for expenses on sale and purchase of home. A further 79 per cent paid a disturbance allowance and 64 per cent provided bridging finance (Merrill Lynch, 1984). The existence of these packages means that the costs of moving have been steadily transferred from the individual to the company. Certainly, compared with the independent migrant, the company mover is very much better off, although the bulk of assistance is directed towards the owner occupier.

The growth and increasing comprehensiveness of assistance packages has been necessary to counter reluctance to move. However, the recent stagnation of the housing market in some regions compared with boom elsewhere has led to concern about the open-endedness of some provisions in the package, most notably the bridging loan. Some tightening-up has occurred, and package reviews brought forward: one company, for example, now finances a bridging loan only after an exchange of contracts has occurred.

Amounts paid vary with circumstances. The key very often is generosity in discretionary payments, especially in the transfer of particular individuals. With major relocations involving all grades of staff there is usually a desire for equality of assistance as far as possible, so more formal provisions apply. Data from the author's survey are available on estimated costs per transfer for 32 organizations in the early 1980s. These show that in 21 cases the cost was less than £10,000 (in 7 cases it was mostly under £5,000), in 4 it was between £10–15,000 and in 7 over £15,000. On these figures the average cost is thus around £10,000 but the median is rather lower – £7,000.

There are many implications of these figures, but three may be mentioned briefly. First, the cost to the company of transferring someone can be very high, so relocation is looked at hard. The highest figures above were almost all inflated by bridging loan interest, and this is an area of assistance that is likely to fluctuate with the state of the housing market – both nationally and locally. Second, the mover is obviously relieved of a major financial headache. Not only are the basic removal and house/sale/purchase costs paid, but the availability of bridging finance, easily arranged and at no cost to the individual, must remove a big deterrent to mobility. Third, the system of assistance that has developed is clearly of major importance but, although some details of packages are known, there has been little systematic analysis of their provisions. It is commonly felt that transferees gain – and certainly do not lose – financially. But is this really the case?

Hourly paid employees and company transfers

For the most part manual grades are recruited and laid off locally. Only in exceptional circumstances, apart from some nationalized concerns, are they required or requested to move.

There are, of course, fundamental differences in attitudes towards careers between manual workers and the more highly paid. A management or professional employee may gain a good deal of interest and satisfaction from work itself, and if not, may anticipate more interest and satisfaction in a new position. The move itself may be welcomed as symbolizing upward mobility and success. If migration is presented as an alternative to redundancy it may be regarded as at least allowing some position in a career hierarchy to be maintained.

For manual workers the major sources of life satisfaction are not, for many, in the work itself, but are more likely to be in the social environment of work and outside work, and this is likely to be even more so for wives. Any move, therefore, is likely to mean very significant losses for little or no gain.

To the lack of jobs presently hindering potential migration for manual workers is thus likely to be added a built-in reluctance to move for reasons connected with work. Not for them longer term career satisfaction. As suggested above, classical and neo-classical economic theory would argue that all workers, faced with redundancy would consider moving to a new location. But how far are manual workers, when presented with opportunities for transfer to another location by their employer, willing to accept a move as an alternative to redundancy?

Analysis of 79 cases from my survey in which the organization had, in recent years (mostly in the 1980s) engaged in closure, relocation or reorganization of units was carried out. Twenty-six cases of major relocation, 48 closures or major rundowns (normally of factories) and 5 major reorganizations had occurred, each of which had presented opportunities for movement for some or all staff. In most cases some residential movement was involved, in other redeployment was to nearby sites within commuting distance. In 17 of the 26 locations, 24 of the 48 closures and all 5 of the reorganizations manual and clerical/secretarial employees were offered moves – usually involving residential relocation – elsewhere. Despite generous financial assistance – and packages were usually common to all grades – all but a tiny handful of manual and clerical staff refused to move. For example, one manufacturer, with 600 redundancies (mainly manual) at a site on the South-east coast, offered to relocate everyone, but only 5 manual workers accepted. In another case a metal manufacturer, closing a plant in Sheffield, offered transfers to Manchester to nearly 200 hourly paid and clerical workers, only 30 of whom accepted. It is often argued that housing is a barrier to movement for the hourly paid. However, several organizations negotiated council housing in the new area for transferees – usually up to a dozen houses – but in no case was the full allocation taken up. Thus, it seems that even when provided with a job with the same company in a new location (thus usually maintaining seniority), and with financial and other assistance to move – even with housing provided – most workers were reluctant to move. Redundancy *in situ* was the preferred option. The major exceptions seem to be nationalized industries where relocation of the hourly paid has long been an accepted and formally negotiated practice.

Government mobility schemes in Britain

Despite its verbal commitment early in its first term of office to labour mobility the present government has done little to foster movement. Indeed, it has presided over the decline and abolition in 1986 of its principal scheme to help the unemployed to move. Measures to facilitate labour migration barely feature in the Manpower Service Commission's programme. The consequence is that those who have positions of strength in the housing and labour markets and can take advantage of private transfer schemes, have found mobility relatively easier while those who relied on government schemes have found it has become more difficult.

Employment transfer scheme

The main public policy vehicle for moving the unemployed into jobs in other parts of the country has been the ETS. Until its abolition in 1986 this scheme provided financial assistance to cover travel and removal costs. A family moving under the scheme could expect a maximum grant of around £3,000.

During the 1970s the ETS developed into a significant element in the total migration, but since the peak year of 1977/8, when 26,539 moved, there has been a steep fall, to 3,299 in 1984/5, the last full year of operation. To some extent this reflects fewer jobs available; perhaps more important, though, has been a general tightening up of the rules governing the scheme since 1979 with a view to achieving greater cost effectiveness. Assistance has always been restricted to unemployed people and those under threat of redundancy, but in the past certain rules were waived for people moving from assisted areas. With the general increase in unemployment and the reduction in public funds available, the rules were applied equally in all areas. Of particular significance in this respect is Rule 4: 'The employer must be able to show that he has tried and failed to attract suitable local unemployed people.' More stringent application of this markedly reduced the possibilities for movement, and this to a large extent explains the reduction in the proportion of ETS movers coming to London (down from 23 per cent in 1978/9 to 12.2 per cent in 1984/5).

Unfortunately, there are no data to allow monitoring of the types of labour being moved. Indications are, however, that latterly the scheme was more concerned with encouraging unemployed skilled workers to move to areas of skill shortage rather than putting unemployed people generally into jobs. Moreover, it performed that task relatively cheaply at little more than £1,000 per person/job. However, the ETS had no housing component and it was up to the individual to make his own arrangements. For some recourse could be had to the National Mobility Scheme, though there are no data on numbers helped by both ETS and NMS.

National mobility scheme

This scheme, which continues to operate, was set up in 1981 to help people move from one area of the country to another. Those eligible are tenants of participating local authorities, new towns and housing associations, people high on their housing waiting lists and others with a pressing need to move. They need to have sound employment or social reasons to move to somewhere beyond reasonable daily travelling distance from their present homes. A participating housing authority agrees to put into a pool at least 1 per cent of its estimated net lettings for the following year for incoming nominees, together with one more letting for each of its own nominees rehoused elsewhere through the scheme. A potential nominee approaches his local housing authority which, if there is a reasonable prospect of a move, makes a nomination to the area to which the nominee wishes to move. There are now 471 housing authorities (including 110 housing associations, and most local authorities, in the scheme).

Since the scheme began 24,032 households have moved with its aid; 18,824 between and 5,208 within counties. Numbers of intercounty moves have steadily increased each year, from 2,737 in 1981/2 peaking at 6,111 in 1984/5 but falling back to 5,607 in 1985/6. The last figure needs to be viewed in the context of 41,387 nominations for the year. Of successful nominees in 1985/6, 28 per cent moved primarily for job reasons, the rest for a range of 'social' reasons, mainly concerned with providing support, for example, 32 per cent were elderly or handicapped people moving to receive support from relatives and friends (National Mobility Office, 1986). The geography of the scheme shows a net positive demand for moves to southern and eastern England, losses elsewhere, and the actual pattern of movement confirms this trend.

The scheme does provide valuable help to those who are successful, but is clearly vastly oversubscribed. Its 'social' nature means it has little to offer as a stimulant for labour migration.

Housing

Thus far this chapter has demonstrated that managerial, professional and skilled technical workers form the bulk of labour migrants, and that a majority of these move within the corporate internal labour market, cushioned from housing market realities by generous relocation assistance. For lower skilled workers, mobility schemes, either private or public offer only limited help. However, the strength of housing barriers is by no means clear and their nature varies with tenure and income. Most manual workers offered a corporate transfer refused, most successful NMS applicants do not move primarily for job reasons.

Much of the discussion about housing barriers to labour migration generally has hinged on the question of tenure. In general owner occupation, private–rented furnished tenancies and tied accommodation have been associated with higher degrees of labour migration, council housing and private rent unfurnished tenancies with less (for example, Cullingworth, 1969; Johnson, Salt and Wood, 1974; Hughes and McCormick, 1981). Prevailing housing tenure trends can thus have a significant role in influencing movement.

The relationship between housing tenure and mobility rate has been documented for over twenty years. Cullingworth's

(1969) study of mobility during 1960–2 showed marked variations between tenure types, with significantly higher rates of movement in the privately rented furnished sector than in others. A decade later this finding was echoed by Johnson, Salt and Wood's (1974) study of labour migrants who also found that the private rented furnished sector had a major lubricating role in providing temporary accommodation while a permanent dwelling in the new area was being sought. Subsequently the National Dwelling and Housing Survey of 1977 and the Labour Force Survey for 1980–1 both indicated the importance of the private rented sector in aiding mobility.

Distance moved and motivation also seem to be related to tenure. Cullingworth (1965) and Johnson, Salt and Wood (1974) found that council tenants moved shorter distances and were less likely to move for employment reasons than other groups. Using census data Minford, Peel and Ashton (1987) showed the importance of the private rented sector for those moving long distances and seeking work; in contrast council tenants predominantly moved locally. The National Movers Survey of 1977–8 (OPCS, 1983) painted a similar picture; Hughes and McCormick (1984) concluded that council tenancy inhibited mobility through its effect on the search process.

The effect of tenure legislation has been discussed by many commentators. Rent acts have made access more difficult to the diminish ing privately rented sector particularly for migrants. It has been argued that housing policies have drastically reduced the migration rates of a large segment of the population (Cullingworth, 1965; Hughes and McCormick, 1981) referring to a 'new law of settlement' being created by the residential qualifications imposed by local authorities. The polemical study of Minford, Peel and Ashton (1987) concludes that falling mobility rates from the middle 1970s can ultimately be attributed to the 1974 Rent Act. However, their partial analysis does not seem to take into account the effect on mobility of economic recession. There is some evidence that the phenomenon of 'discouraged' movers occurs. The General Household Survey for 1976 suggested that 3.1 per cent of all households in England and Wales contained people who had seriously considered moving but had either given up or changed their minds; housing reasons being the main stumbling block. The National Consumer Council's survey (1984) indicated that it was in the local authority sector that discouragement was highest, suggesting that it was not necessarily the rent acts that prevented people from moving.

Underlying the tenure-mobility relationship has been a sea-change in the balance of tenure types. From 1971-85 owner occupation went up from 49 per cent to 61 per cent of households;

council renting rose during the 1970s from 31 per cent to 34 per cent in 1981, but had slipped back to 28 per cent by 1985 in response to sales of council housing; private rented accommodation halved, to 7 per cent. The overall effect of these changes should be to ease the situation for better-off migrants while making it more difficult for others. There is as yet little empirical evidence of the extent to which this shift in housing tenure has differentially affected mobility. Congdon and Champion (1987) have concluded that immigration to the South-east has not been in response to increases in owner-occupied stock there, and that new building has had only local effects.

Although owner occupation has been associated with greater mobility, recent trends in house prices across the country may be changing that relationship in a strongly geographical way. The regional rank order of house prices has remained fairly constant in recent decades but in the last ten years or so the gap between top and bottom has widened (Nationwide Building Society, 1987). Of particular importance has been the growing disparity, especially since 1984, between South-eastern England (especially London) and the rest of the country. For example, in 1975 prices in the capital averaged 57 per cent higher than in the northern region; by 1987 they were 142 per cent higher. There is growing evidence of the ripple effect of rising house prices extending beyond the South-east, often, it seems, following the motorways and upgraded railway lines.

One way of assessing the degree of difficulty facing the labour migrant moving to a higher priced region is analysis of the ratio between house prices and incomes. In South-east England in 1986 the average household would have had to find 5.2 times its annual income to buy the average house, compared with less than 10 per cent in most northern and western regions. With an increasing proportion of the national population living in owner-occupied property, house price barriers may now provide the biggest set of obstacles to labour mobility.

Ultimately, the limiting factor on house price increases is people's ability to buy their way on to the escalator in the first place. There is evidence that past relationships between property prices and incomes may be changing. Partly this is due to inherited wealth as the first big generation of owner occupiers dies. Others are able to step on to the escalator in company, perhaps two or three people clubbing together to put down a deposit and share a mortgage. The March 1988 budget, effectively reducing tax allowances on mortgages where such co-operation occurs, may ultimately have a braking effect. This already seems to have happened in Greater London, where in the first quarter of 1988 average house prices failed to increase. Though, in the late 1980s mortgage finance is more freely available and from a wider range of sources than in the past. Home ownership, therefore, continues to grow, especially among young professionals who are in occupations which are geographically more mobile anyway.

Rises in house prices and growing regional differences affect labour migrants in several ways. They have an inhibiting effect on owner occupation, especially for young people, and they also lead to price gradients too steep for many to climb. One response to this situation has been the growth of long distance commuting on a Monday–Friday basis; one report in 1987 showed ten to fifteen long distance commuters on every North-east bound coach leaving London on a Friday evening (Policy Studies Institute, 1987).

As house prices rise there is a knock-on effect for the costs of moving since agency and legal fees and stamp duty all change in line with house value. It is thus more expensive to move house within or into southern England. People in the best position are those for whom removal costs are paid, usually by the employers, but there are signs, still largely anecdotal, that this practice is beginning to create its own migration fields in the South-east and East Anglia as companies prefer to recruit only from potential staff already living in the high price area, thus minimizing their outlay on relocation costs.

It is now doubtful that the government is serious about encouraging the unemployed to move into jobs elsewhere. There is no sign of the sorts of policies that would be necessary to help low income migrants, such as the abolition of length of residence qualifications for council housing, the development of comprehensive means of information exchange in the council sector (perhaps from a central clearing house, so that tenants can more easily transfer between local authorities) and an increased provision by local authorities of a stock of accommodation specifically for allocation to migrants. Where the private-rented sector is small and under severe pressure greater flexibility in the allocation of council housing is essential. There is no doubt that for migrants a vibrant private-rented sector, at the inexpensive end of the market, is necessary. While the need for tenants' rights must be safeguarded, the trend of the Rent Acts of the last thirty years must be reversed. The polarized housing market we are now moving towards is too inflexible. There is as yet little evidence that proposals 'freeing' the private-rented sector will be effective, but in any case market rents will be out of reach of many on low incomes. Housing policies also need to be considered in the light of local conditions. Some places are more expensive to move to than others, so that levels of assistance ought to vary according to destination.

Conclusion

We can no longer assume, if we ever could, that labour migration will alleviate the problems of structural employment. In a highly segmented labour market, with a growing dependence on skill and expertise, labour migration is now more concerned with placing individuals who already have jobs into vacancies elsewhere. Most of them will already have the new job before moving, they will not migrate speculatively. There are some, however, again unquantified but undoubtedly including the young and single, who might move in search of work. For them the lack of cheap temporary accommodation can be a major headache.

In general, one has to say that there is no clear evidence that housing *on its own* inhibits labour migration; it is one type of barrier among several and the ability to surmount these barriers depends on the resources, especially of finance, information and time, available to potential migrants. The role of housing in labour migration remains elusive, for 'housing' can be several things for a migrant, and the nature of housing problems confronting the potential mover varies with the individual: 'strength' in the owner-occupied sector is not the same as that in council housing. In 1987 the prime constraints on labour migration should not be sought only in the housing market but in the operation of the labour market where the provision of information, training and work organization can help fit people into vacancies, and in the attitudes of individual households towards moving to another part of the country.

Doing anything about these major problem areas requires substantial policy initiatives which are unlikely with present high levels of unemployment and government fiscal policy. Nevertheless, it is possible to identify significant issues. In particular there is a need to develop housing and migration policies that are more compatible than exist at present. Over the years government involvement in the housing market has steadily increased. A significant result has been to restrict entry by labour migrants into the rented sector. Yet, at the same time, council and private rented housing is that which is most likely to be provided for those workers (the lower paid and unemployed) at whom the ETS was aimed. This contradiction, as well as the lack of government commitment to a migration policy, means that the pattern of labour migration that exists is very much the result of *laissez faire* policy and has to find its own adjustment to the constraints imposed by housing.

References

Atkinson, J. (1987), *Relocation News*, 4, pp. 17–18.

Bonnar, D. M. (1979), 'Migration in the South-east of England: an analysis of the interrelationships of housing, socio-economic status and labour demand', *Regional Studies*, 13 (4), pp. 345–60.

Congdon, P. and Champion, A. (1987), 'The migration turnround in London and its relationship to employment and housing markets', (unpublished).

Conway, J. and Ramsay, E. (1986), *A Job to Move* (London, SHAC).

Cullingworth, J. B. (1969), *Housing and Labour Mobility: a Preliminary Report* (Paris, OECD).

Flowerdew, R. T. N. (1976), 'Search strategies and stopping rules in residential mobility', *Institute of British Geographers, Transactions*, New Series, 1 (1), pp. 47–57.

Gleave, D. and Palmer, D. (1980), 'The relationship between geographic and occupational mobility in the context of regional economic growth', in J. Hobcraft and P. H. Rees, *Regional Demographic Development* (London, Croom Helm), pp. 188–210.

Gordon, I. (1979), 'The analysis of motivation specific migration streams', Urban and Regional Studies Unit, University of Kent (mimeo).

Gordon, I. and Vickerman, R. (1979), 'Opportunity, preference and constraint: an approach to the analysis of migration in the London region', Urban and Regional Studies Unit, University of Kent.

Green, A. E., *et al.* (1985), 'What contribution can labour migration make to reducing unemployment?', *Discussion Paper* 73, CURDS, University of Newcastle upon Tyne.

Hamnett, C. (1984), 'Housing the two nations: socio-tenurial polarisation in England and Wales', *Urban Studies* 43, pp. 389–405.

Hughes, G. and McCormick, B. (1981), 'Do council housing policies reduce migration between regions?', *Economic Journal*, 91, pp. 919–37.

Hughes, G. and McCormick, B. (1984), 'Migration intentions in the U.K.: which households want to migrate and which succeed?', *Economic Journal* (Supplement), 95, pp. 113–23.

Jackson, J. A. (ed.) (1969), *Migration* (Cambridge, Cambridge University Press), Sociological Studies 2.

Jansen, C. (1969), 'Some sociological aspects of migration', in J. A. Jackson (ed.), *Migration* (Cambridge, Cambridge University Press), pp. 60–73.

Johnson, J. H. and Salt, J. (1980), 'Employment transfer policies in Great Britain', *Three Banks Review*, 26, pp. 18–39.
Johnson, J. H., Salt, J. and Wood, P. A. (1974), *Housing and the Migration of Labour in England and Wales* (Farnborough, Saxon House).
Johnson, J. H., Salt, J. and Wood, P. A. (1975), 'Housing and the geographical mobility of labour in England and Wales: some theoretical considerations', in L. A. Kosinski and R. M. Prothero (eds), *People on the Move* (London, Methuen), pp. 91–101.

Lind, H. (1969), 'Internal migration in Britain', in J. A. Jackson (ed.), *Migration* (Cambridge, Cambridge University Press), pp. 74–98.

McKay, J. and Whitelaw, J. (1977), 'The role of large private and government organizations in generating flows of interregional migrants: the case of Australia', *Economic Geography*, 53, pp. 28–44.
Mangolam, J. J. and Schwarzweller, H. K. (1968), 'General theory in the study of migration: current needs and difficulties', *International Migration Review*, 3, pp. 3–18.
Massey, D. (1984), *Spatial Divisions of Labour* (London, Macmillan).
Merrill Lynch Relocation Management International (1984), *Employee Relocation Policies Among Major U.K. Companies* (London, Merrill Lynch).
Milkovich, G. T., Anderson, J. C. and Greenhalgh, L. (1976), 'Organization careers: environmental, organizational and individual determinants', in L. Dyer (ed.), *Careers in Organizations* (NY State School of Industrial and Labour Relations, Cornell University).
Minford, P., Peel, M. and Ashton, P. (1987), *The Housing Morass* (London, Institute of Economic Affairs), Hobart Paperbacks 25.

National Consumer Council (1984), *Moving Home* (London, National Consumer Council).
National Mobility Office (1986), *Annual Report* (London, National Mobility Office).
Nationwide Building Society (1987), *House prices: the North/South divide: Who gains – who loses*? (London, Nationwide Building Society).

Office of Population – Censuses and Surveys (1983), *Recently Moving Households – a follow up to the 1978 National Dwelling and Housing Survey* (London, HMSO).
Office of Population – Censuses and Surveys (1986), 'General Household Survey: preliminary results for 1985', OPCS *Monitor*, GHS 86/1 (London, HMSO).

Policy Studies Institute (1987), 'The long distance commuters', *New Society*, May, pp. 11–13.

Salt, J. (1984), 'Labour migration within multi-locational organizations in Britain', *Final Report*, ESRC Grant F/00/23/0027.
Salt, J. (1986). 'Labour force and employment in western Europe', in W. T.

S. Gould and E. Lawton (eds), *Planning for Population Change* (London, Croom Helm), pp. 39–63.

Salt, J. and Flowerdew, R. T. N. (1986), 'Socio-economic selectivity in labour migration'. Paper presented at the Conference on Comparative Population Geography of the UK and Netherlands, Oxford, September.

Sell, R. R. (1983), 'Transferred jobs: a neglected aspect of migration and occupational change', *Work & Occupations* 10 (2), 179–206.

Taylor, R. C. (1969), 'Migration and motivation: a study of determinants and types', in J. A. Jackson (ed.), op cit pp. 99–133.

Tebbit, N. (1981), Speech made at Conservative Party Conference, October, when Secretary of State for Employment. Reported in Green *et al.*, op cit.

Thatcher, M. (1980), Speech quoted in *The Times*, 21 July. Reported in Green, *et al.*, op cit.

Wolpert, J. (1965), 'Behavioural aspects of the decision to migrate'. Papers, Proceedings of the Regional Science Association 15, pp. 159–69.

4 *Housing markets and labour markets in the quartered city*

PETER MARCUSE

Introduction

The relationship between work and housing is much more than a relationship between labour markets and housing markets. The nature of the work *process* – not just how much is paid or where jobs are located – influences the type of housing that is demanded and the type of housing that is supplied. The chapter provides a few examples, and then uses two cases, gentrification and homelessness, to illustrate the contention that the linkage between labour and housing is a fundamental *internal* one, in which the content of labour and the form of housing are directly related, and are *prior* to the external relations of the markets for each. The chapter ends with the suggestion that these linkages vary by class, and suggests as prominent on a future research agenda an examination of the internal relationship between class, work processes and the nature of housing provision. The examples used come from the United States, primarily New York City, but the processes clearly appear in all industrialized private market economies, and many aspects are relevant to Third World and socialist economies also.

The levels of labour–housing market relationships

The relationship between housing and labour markets exists at three levels. They will be referred to as external and internal. The external relationship is concerned with how markets in two commodities, each following their own laws, relate to each other.[1] The internal relationship is concerned with the ways in which the labour processes affect the nature and form of housing, as well as the ways in which the shape of the housing supply affects the nature and location of work. Ray Pahl's recent work (1984) comes the closest to the approach suggested here. The internal relationship is the primary focus of this chapter. But first a word on some of the research issues involving the external relationships.

External labour–housing market relationships

External labour–housing market relations are here defined as those that involve the way in which the exchange – the buying and selling – of commodities in the two markets relate to each other. Most classical economics looks at the relationship this way;[2] so, indeed, does classical Marxist theory.[3] It is a useful form of analysis, subject to limitations discussed below, but it has certain problems even within its own limited framework. Some are well known[4] – the handling of externalities, for instance; others are not. Let me mention only three:

1 We often speak of 'government intervention in the private housing market' as if government action is an external, distorting factor superimposed on a pre-existing private housing market. The term is wrong, and indeed ideologically loaded. Government does not 'intervene' in some previously constituted private market. Government is an essential element in that market from the very outset. A private market in housing cannot exist without roads, sewers, use regulations, height restrictions, courts to enforce contracts, police to prevent trespass, and so on and so forth. Yet most studies of the impact of government on housing market behaviour assume some perfectly functioning market which is then 'distorted' by government action.[5] We need a more sophisticated research approach, one that understands the impacts of government regulations or actions in the context of the impacts that their absence would have, and then builds the necessary actions into the model.

2 Most studies assume a single 'supply' side. But in fact it is multiple and contradictory. Land owners have interests clashing with those of builders; landlords of existing buildings have an interest in the absence of new construction, construction trade unions and financial institutions in its presence; lawyers and brokers profit from the level of activity regardless of direction, investors wish to see activity go only up; and so forth. In various locations and at various times these interests will have varying impact. No theory or research that does not take these differences into account will go far in explaining reality.

3 Most studies of the external relations between labour and housing markets assume proximity to work to be the key labour–market-related characteristic that influences housing demand. This is open to question, for internal reasons, below; but even externally, proximity to work is only one of many factors that influence choices (when there are choices) as to housing, and research as to the relative weight among these factors is still in its conceptual infancy. Most studies are simply empirical: they ask a respondent, 'why did you move?' and then calibrate and scale the answers. But such a procedure weighs apples against oranges, with no basis in a theory of what the relationships are. Will the importance of cost be the same for all income levels? Will long-term cost or short-term cost be decisive (for example, is appreciation reckoned in, and if so with what discount rate)? Will the answer differ by age, by income, by level of prosperity, by security of job? To what extent are preferences moulded by ideology (the status of home owner, for instance[6]) and to what extent do changes in the real context make a difference?

But the major limitation of a focus on the external, market relationships between labour and housing is that it assumes housing is supplied *exclusively* as a *commodity*. A market, after all, is defined as where commodities are bought, sold and exchanged. But housing isn't *only* a commodity. It is to a significant extent provided for reasons not having to do with profit from its production or sale, but rather profit or other benefits to which housing contributes.

How housing is disposed of as a commodity is well known: its price is essentially determined as is that of any other commodity, and if a given worker cannot afford it, he or she makes do with less.[7]

We begin to get into trouble with the sufficiency of the commodity characteristic of housing precisely when we examine its relationship to the labour process. Housing is a commodity that is a necessary element in the reproduction of labour. Producers, therefore, have a collective interest in keeping its price low. Thus, in the nineteenth century, many employers provided housing for their employees directly, not as a commodity, but as part of their wage (Weisser, 1975 pp. 23, 27, 30).[8] The practice is less frequent today, although employer payment of moving costs and subsidization of relocation housing is frequent for executives or 'key' employees when a company relocates. The *New York Times* constantly prints stories in its Real Estate section about how the high cost of housing is causing firms to leave the area, or requiring them to pay differentially high wages. Thus the *Times* joins many in the business community concerned with holding the price of housing down; the banks formed the New York City Housing Partnership to provide 'non-profit' – read, 'partially decommodified' – housing for clerical and lower level service workers in the city. The exigencies of the labour market lead to a rejection of the housing market as a mechanism for the provision of housing in key sectors.

Nor does a market analysis based on the commodity form of housing explain such important facets of the housing system as tenure relationships. The prominence of home ownership, the decline of the private rental market,[9] the ups and downs of social housing, have to do with tax policies, with land use controls, with racial and ethnic patterns, with traditions of food production, and, heavily, with ideological presuppositions and goals, whether phrased in terms of Herbert Hoover's view of 'the American Home . . . [as] the foundation of American life', or Franklin Roosevelt's less rhetorical 'homeownership . . . a guarantee of social and economic stability', or indeed Engels' more cynical comment:

> Give [the propertyless proletarians] their own houses, chain them once again to the soil, and you break their power of resistance to the wage cutting of the factory owners.[10]

Something more than a market linkage between housing supply and labour conditions is required to explain patterns such as those of tenure in housing.

The internal housing–labour process link

The internal linkage between labour processes and the provision and use of housing runs underneath the market, and is a precondition of the interplay of market forces (and their political manifestations as well). The link between work processes and housing produces results with which the market, and the commodity form in which housing appears in that market, have nothing to do. The motivations of both providers and users of housing are shaped by the labour process: the market then provides some of the field on which these motivations are played out, constrained, and expressed; but only some, not all, of that field.

Some short examples, and then two current cases at more length:

1 *Suburbanization* is often presented as resulting from the market satisfaction of simple consumer preferences for a particular style of living, preferences newly capable of being met because

technical and business changes enabled developers to respond to these long-standing consumer preferences (Jackson, 1985). The movement of jobs to the suburbs then follows the housing market's lead. The suburban movement of jobs is accentuated by the desire of manufacturing industry for larger surfaces for efficient production, and of white-collar employers for environmental amenities, which increase the lure of the cheaper land to be found in the suburbs. So changes in labour markets and in housing markets begin independently of each other, each driven by their own logic, but adapt to each other in an external manner as they develop.

In contrast, a strong argument can be made that links suburbanization to the attempt by employers to escape social factors present in concentrated fashion in the city: labour militancy,[11] competitive availability of jobs, labour 'unreliability'. This argument also explains suburban housing patterns as adapted to the specific forms of work required in the suburbanized factories, and later offices.[12] Jobs paying the lowest wages could not move out – those holding them were dependent on the cheap if substandard housing of the inner cities. Top decision-making jobs could not move out – decision-makers wanted to remain in direct contact with their peers, and centralized control and centralized location went together. But suburban housing patterns meshed perfectly with the sessile but skilled workforce needed for middle level management, white collar, and better paying production jobs. The labour market and the housing market meshed to satisfy these internally linked needs.

2 *Anti-displacement sentiment*, the unwillingness of residents to move, is handled in conventional economics as a transaction cost, a friction in the housing market. In part, this is appropriate. Anti-displacement feeling, however, is not felt equally by all people at all places at all times, nor is it simply a matter of the financial costs of moving or even the costs of adjusting to a new environment. In the 1920s in the Jewish working-class quarters of New York City, 1 September was a traditional moving day. If a family were successful, it would be to a bigger apartment; in bad times, it would be to a cheaper apartment. The indicator of real success was moving out of the old neighbourhood, perhaps to the newer Art Deco buildings on the Grand Concourse in the Bronx; if success were headier, to the suburbs. The strong sense of neighbourhood, the resistance to being dislodged, is not to be found in the historical record of the times. And whether or not it existed for European immigrants to the USA or foreign workers going to more prosperous European countries, it was

not powerful enough to prevent their moves. It is the alternatives that count, and the weighing is more often between job and community than it is between one community and another.

One hypothesis to explain the strength of anti-displacement sentiment among workers and the poor today relates the resistance to moving directly to changes in the labour market. In the 1920s upward mobility was the expectation; today, safety in an existing position is, for many, the only aspiration, and upward movement hardly a hope. Security, stability become virtues in themselves. For the poor, part of the basis of survival lies in the informal networks of support (and income, in the informal economy) that are neighbourhood-based.[13] The resistance to moving may still be called a transactional cost; but its magnitude is determined by the characteristics of the labour market, not the housing market.[14]

3 A final example: *homework by professionals*. The number of intellect-workers[15] has increased rapidly in this century. For many, communication and control aspects of their jobs require a location close to the centres of business and power; gentrified housing has major appeal for them, as is argued below. For others, given the nature of their work and advances in technology, location is of little relevance to the performance of their tasks. Exurban locations and homes (wherever located) with space for their jobs as well as their residences become possible.

The argument that new technology permits more who always wished to work at home to do so is well known. But the reality seems to lag far behind the potential. In theory, homework permits better integration of work life and home life. That may be true in fact for artists, or some intellectuals; it certainly was not true of the garment workers who took piecework home to do in their tenements, nor is sitting before a computer at home much more integrated with personal life than doing the same thing downtown. For exurbanites, less frequent but longer trips to the employer's office may result in the same net travel time as more frequent but shorter trips might have meant.

Beyond that, there are pushes for more intellect-work at home that come not from the personal preference of the worker, but the benefit of the employer. Homework has major advantages for the employer, given the changing nature of the work process. It permits the use of a contracting-out system, rather than a salary system: intellect-workers can be put on piece work, if they work at home in their own hours and have only to turn in a product at the end. They can be laid off easily, their compensation can be lowered by competitive bidding or 'sourcing out', the danger of their acting in concert is reduced,[16] the costs of providing a workplace can be passed from employer to employee, the rewards of self-motivation can be more easily achieved, the necessity of company supervision dispensed with. The 'free' rewards of an environment of personal choice, be it countryside or loft, can benefit the employer as well as the worker. That possibility promotes the spread of forms of housing not made possible by changes in the labour market or housing market, but rather by changes in the labour process.

So the argument here is that the nature of the labour process, as well as, and perhaps even more than, the negotiations that take place in the labour market, shapes the provision of housing, including that part of the provision that runs through the conventional housing market as well as that part that runs through the state. Two current examples highlight the point.

Gentrification and the labour process

Gentrification[17] is frequently seen as the result of a change of tastes: a re-awakening to the virtues of inner city living, a rediscovery of the charms of historically preserved older buildings, a swing of the pendulum of consumer preferences from the earlier suburban momentum. Why the pendulum should swing just now, and why in this particular direction, is not explained.

Gentrification and the specific form of housing preferences (housing demand) it represents are not random shifts of preference, it is argued here, but occur at a specific place and time because of specific changes in the labour processes for a specific group/class of people. In market terms, the processes are fairly clear: a reduced demand for unskilled labour reduces the effective demand for cheap housing near the centre of the city; an increased demand from professional/managerial employees for central city housing and increased costs of transport and congestion inevitably result in displacement. But the new demand is for a specific type of housing, a specific type of neighbourhood. The argument, in a nutshell, is that gentrified living (living like the gentry, a major component of which is living in gentrified neighbourhoods) provides, for those engaged in work demanding creative but inherently unproductive and unrewarding labour, a set of satisfactions not available from their work but necessary for their motivation. The argument is in a sense the negative of Louis Wirth's: rather than seeing specific forms of urbanizm as a reaction to life in cities, it suggests that the quest for specific forms of (urban) living defines specific types of neighbourhoods in cities (Wirth, 1938, pp. 1–24).

Economic (industrial) restructuring has, in the postwar period, created the need for an expanded, technically competent, and strongly motivated work force. But the traditional mechanisms for producing that motivation, for obtaining the acquiescence and support of those doing such work, no longer function. Too many of these new workers are no longer pulled by the lure of entrepreneurship, power and riches,[18] nor pushed by the fear of penury and hardship. The stratum in question involves neither the idle rich nor the decisive powerful, but the well paid administrative, managerial, technical culture, professional workers, often with inherited as well as current income. Gentrification can be seen as one way of dealing with this problem of motivation, acquiescence, satisfaction.

The classic pattern of gentrified housing is the converted brownstone house or reconstructed old building, expensively updated but with conspicuous symbols of its past left exposed, with elaborate cooking and entertainment 'centres', furniture consciously chosen for aesthetic qualities, often by a professional interior designer, all used to display acquisitions and reveal character. The gentrified neighbourhood is replete with fern bars, plants in windows, small tables next to the pavement where those inside can be seen from the outside, and a line waiting to get in to pay high prices for limited food but an unlimited sense of being 'in'. The gentrification of people (Rose, 1984) is happening through the evolution of a sophisticated consumerism, not only as to goods but of culture, the environment, and, crucially, of residential space. This gentrification is thus a mechanism to provide a motivation for and satisfaction with continued integration into unproductive and humanly meaningless jobs, where quality of performance is important but usefulness of result is not. The gentrification of space is both a by-product of, and contributes further to, this process.

Not all gentrifiers are alike. Theories of neighbourhood change conventionally distinguish between the 'pioneers' and the 'settlers'. The pioneers are younger, take greater risks, do substantial rehabilitation work themselves, and make handsome profits when they leave. The settlers have higher incomes, are concerned about the safety of the neighbourhood, and are willing to pay higher prices to move into a neighbourhood others have already gentrified. Both groups fit into the widely used category of 'yuppie'; they are insiders, doing well in the system, know how to manipulate it and find new residential patterns that provide them with intangible as well as tangible rewards: excitement as well as profits, status as well as shelter.

Damaris Rose uses the same term, 'pioneers' to speak of what seems to be a separate although related phenomenon. They are often 'educated young people . . . find[ing] creative ways to respond to new conditions of paid and unpaid work and worsening economic conditions' (Rose, 1984). Those involved are seeking an alternative lifestyle, alternative forms of housing and ways of earning a living; they are trying to escape from or avoid the system, not manipulate it to ends identical with those having more conventional jobs and homes. The pattern certainly exists, and is much more visible in Europe than North America. The 'alternative scene' in cities such as Berlin, Frankfurt, Amsterdam and Copenhagen includes many who are seeking inexpensive housing in patterns of collective occupancy, as a way to avoid rather than profit from current residential trends. The converted factories or farm buildings or 'slums' they occupy may or may not in the future be desirable for others of higher income and greater purchasing power; that is not their concern. Their housing patterns may appear similar to those of other gentrifiers, but their work and role in the labour process is very different.

The demand for the specific form of housing that gentrified neighbourhoods and gentrified houses represent thus is a response to changes in the nature of work for a particular group; changes that create a need which the new form of housing fulfils. The need is not one that arises out of some unpredictable and arbitrary swing of popular taste, nor is it solely an economic response in housing to changes in wage patterns or job location in the labour market. Rather, gentrification, in its specific social, aesthetic, and cultural forms, is internally linked to changes in the work processes in which gentrifiers are engaged.

The pattern may be likened to that prevalent during a major part of the suburbanization boom, and is very much class-specific. The single-family house (or its essential form, sometimes duplex or row house construction) with its own plot of land, its own mortgage, its own responsibilities for separate care, maintenance improvement and enlargement, its broad freedom of use, its presumed stable market value and possibilities of appreciation, represented a sharp social and cultural rupture with the traditions of working-class quarters in old sections of the city, on the one side, and, on the other side, an equal rupture with the traditions of 'urbanity', cosmopolitanism, political engagement with strong components of independence and radicalism. Suburban housing was specifically adapted to meet the life situation created by a specific position in the work process: significant material income, limited security, limited creativity, immediate rewards but the lingering concern about discontent and resistance to workplace discipline and workplace compensation.[19]

The earlier workplace disciplines needed in the sweat-shop and industrial jobs of the nineteenth century can similarly be seen reflected in the housing provisions considered most desirable for the holders of such jobs. The moulding of character and creation of incentives to work that Octavia Hill tried to accomplish in the tenement houses of London, or those of the philanthropic reformers in New York City at the turn of the century, play the same role, if for a different class, that the suburbs play for skilled workers thereafter and that gentrification plays for the professional/managerial class today.

The 'housing' of the homeless

The 'housing' of the homeless seems initially determined completely by the housing market, by the commodity character of housing: no money, then no housing. But no money is in turn determined by the labour market: no job, then no money. The 'reserve army of the unemployed' is no longer needed for work near the city centre.

Others, however, have a new need to be very close to the central city. Long hours of heavily interactive and specialized work must be done at exceptional concentrations of work. So there is mounting demand for nearby residences. 'Waste' – cheap – residential space near the new concentrations of labour is uneconomic in the market-place. So the poor are ousted from flop-houses, single-room-occupancy buildings, the cheap hotels of skid row, near the downtown. Their land has become valuable again. The ultimate cause of homelessness is not speculation, but changes in the organization of labour that call for new degrees of spatial concentration of work and of control. So on the surface, homelessness is a classic product of the link between labour and housing markets, and its source must indeed be looked for in the changing structure of the labour market. But the roots of the peculiar treatment of the homeless go deeper (Marcuse, 1988b).

Both the former housing of the homeless and their current non-housing quarters, whether in a bus station, railroad terminal, subway car, public building, or private office doorway, are incompatible with the content of the work being done downtown. That work no longer 'produces' anything other than money; it is entirely devoted to increasing control, increasing the appetite for consumption, centralizing power, claiming ownership and influence. The poverty of the homeless challenges the pursuit of wealth as an end in itself; greed becomes less acceptable in the presence of poverty. The commuters coming into the bus stations, the train stations, the airports, bent with the strongest motivations on the pursuit of ignoble ends, are afraid of the implicit challenge to their values. The challenge to conscience results in an unprecedented upsurge of charitable actions by established philanthropies and individuals, but it also results in an unprecedented callousness towards human beings in a 'civilized' society. When leaving open the doors to the entrance ramps at Grand Central station in New York did not make it cold enough to force the homeless sleeping there to move, the concrete ramps were washed down with lye to create so strong a smell that no one could possibly remain huddled there (Kozol, 1988). The challenge to conscience is more than a charitable gesture or two can dispose of.

Thus the disturbing sight of the homeless results in major efforts to relocate, rather than rehouse, the homeless. Out of sight of the business men and women, the commuters and visitors to the city, is the *sine qua non*. But where the homeless can then be put is severely limited. Some, of course, don't care:

> If it is illegal . . . to litter the streets, frankly it ought to be illegal for people . . . to sleep in the streets. Therefore, there is a simple matter of public order and hygiene in getting these people somewhere else. Not arrest them, but move them off somewhere where they are simply out of sight. (George Will, in Marcuse, 1988a, p. 70).

For those slightly more enlightened, finding locations for the homeless in the more removed working-class areas of the city (higher income neighbourhoods are too expensive for such purposes) creates its own set of problems, again as an outgrowth of developments in the field of work. The solidarity that once existed between the poor employed and the poor unemployed has been broken. The unemployed are seen as a threat by many of those who are themselves insecure in their employment or status. The 'threat' of the homeless is both in the labour market – where the traditional fear is that their competition will reduce wage rates – and in the housing market, where the fear is that they will lower 'property values' and interfere with self-images of 'good' neighbourhoods. Hence the 'NIMBY' phenomenon: residents who say, 'shelters perhaps, but Not In My Back Yard; somewhere else'.

The explanation for the present state of the homeless lies at two levels. On the market level, market changes in the value of the space that they used for housing leads to their displacement; that change in value is in turn produced by changes in the labour market and the demand for certain types of work at high levels of spatial concentration. At a deeper level, the housing that those now homeless get (if any) is dictated by changes in the type of work being performed in the society: less of what they formerly did, and much more of a kind of work psychologically and socially incompatible with what they are expected to be able to do. The gentrifiers who displace the homeless are disturbed by them, and want them out of sight. The very existence of homeless people is discordant to the professionals, technicians, managers; something is being managed wrongly, somewhere. If the reaction were as it was in the 1960s, with significant sections of those who should be doing the intellect-work that keeps the economic wheels turning opting out of the system, or protesting its inequities, there is no telling what might happen. If those still employed at more traditional wage work see the interests of the homeless as linked to their own, there is no telling what might happen. Better to remove the homeless from sight, warehouse them in shelters where they can do no harm to anyone but themselves, or keep them out of town altogether, and solve the problem that way.

The market displaces the poor from their former housing; the values and insecurities produced by the work process in those better off than they, keep the poor, once homeless, from getting the homes they need, where they need them.

The magnitude of the issue has not gone unnoticed. Mayor Edward Koch of New York City, testifying before a Congressional Committee in Washington in December 1987, accused advocates for the homeless of 'trying to bring down the system' by demanding decent housing, not just temporary shelters, for the homeless. The 'system' in question is both the housing system and the labour system.

The quartered city: the class–specific relations between housing and work

This leads to a final point: the relationship between housing/labour markets and class.

'Dual city', 'polarized society', 'increasing distance between the rich and the poor', are phrases used repeatedly to characterize contemporary developments; they all miss the specific historical changes taking place today. Much more useful would be a theory of a four-class, not a two-class, society; a quartered, not a dual, city: upper class (dominant, ruling, capitalist), professional/managerial class (yuppie, technocratic, upper middle), working class (blue collar, white collar, skilled–unskilled), very poor class (partly and under-employed, informal economy, marginalized, state-dependent). These are tentative, loose formulations, intended to highlight work-based divisions as reflected in housing and urban patterns,[20] rather than as rigorous production-based definitions of class.[21]

To the extent that housing research has related itself to class categories, the focus has been (somewhat surprisingly) on tenure differences. Phrases such as 'tenure-based social polarization' and 'widening gap between those at either end of the income hierarchy' are used to describe changes in tenure (Hamnett and Randolph, 1986). Yet it is clear that more than two distinct groups are involved, and that within each tenure category persons with various positions in the labour force (or outside it) have a variety of accommodations very different from each other in location, in quality, price, size, security of occupancy, etc. The picture of a two-tenure-class society certainly fails if the suggestion is that private rental tenure and home-ownership parallel labour categories; it fails as well when social housing is brought into the picture, either in place of private rental (as is increasingly the case in Britain) or in addition to it. Even in the USA, where public housing is almost exclusively occupied by those at the most marginalized end of the labour market, public housing occupancy does not parallel any production-based class category: a third of New York City's public housing residents are regularly and securely employed, frequently in the public sector, and nationally only a small fraction of those eligible for public housing by income are housed in that tenure (Dolbeare, 1985; Marcuse, 1988b). The tenure divisions of the housing market are not congruent with the class divisions of the labour market.

Other definitions of housing differences may be more useful to illuminate the relationship between production-defined classes and housing-defined cleavages. Nor are housing classes adequate to explain political behaviour or life style or economic position. Tenure is more a result than a cause, and cleavages within tenures are for most purposes more important than cleavages among tenure categories; more research needs to be done to work out their precise relationships.

Conclusion

Two conclusions seem to me to flow from this analysis: one objective and research-orientated, the other more subjective and more political.

From a research point of view, the nexus between work and housing should be a high priority issue. The linkages between the two are central to each; if their relationship is not understood, neither of them will be understood separately. And research on that nexus needs to develop the analysis of external labour market-housing market linkages – a few areas that need refinement are suggested above, and there are many more – but it needs to go much further. Research needs to look at the internal connections between changes in the processes of work and changes in the forms of housing, what they are, how they influence each other, how they can be influenced by public policy. Some smaller examples and the larger cases of gentrification and homelessness outlined above have tried to show where such research might lead. A narrow economistic view of the linkage between work and housing would be fatal; a broad view, however, should be very productive indeed.

The second, more political, conclusion is as follows: we are, in the United States and in several other developed private market economies, in the midst of a housing crisis that affects large portions of the population, a crisis that, for perhaps the first time since World War II, is getting worse and not better. That crisis is not going to be resolved by minor adjustments in the housing supply system, by tinkering to make the housing market function 'better', by finer calibration of the existing market in housing with the existing market for labour. On the work side, a system that provides more and more people with no jobs at all, at the same time that others are being asked to work harder and harder at more and more challenging but also more and more meaningless and unproductive work, needs a fundamental reassessment. On the housing side, a system that initially distributes housing only as a commodity, and modifies that distribution only when it serves the interests of an equally asocial labour market also needs fundamental reassessment. No campaign to deal with housing problems will come close to real solutions unless it begins to address both sides of the work process/housing supply system linkage.

Notes

1 A third level of analysis requires (but will not receive here) a *comprehensive* examination of the relationship between the two markets as component parts of a larger structure in which many other activities which are part neither of the labour nor of the housing market – those of government, for instance, or those determining interest rates, or those affecting political stability – affect both housing and work, and their relationship to each other.

2 For a sophisticated recent example, see Hughes and McCormick, 1987, pp. 615–45. The useful work of George Sternlieb and his colleagues at the Center for Urban Policy Studies at Rutgers in the USA approaches the issues in similar manner. See, for instance, Sternlieb and Hughes, 1976.

3 In the case of housing, Engels says, 'we are dealing . . . with a quite ordinary commodity transaction between two citizens, and

this transaction proceeds according to the economic laws which govern the sale of commodities in general . . .' Since labour power is similarly bought and sold, 'every permanent price reduction in the worker's necessities of life [such as housing] . . . will . . . finally result in a corresponding fall in wages' (1935). See also Angotti, 1977. Engels (and Angotti) do however consider non-commodity aspects of housing.

4 See, for instance, the research reported in Hoenderdos, Hooimeijer, van Hoorn, 1986.

5 See, for instance, the old studies for the Douglas Commission, or the series of studies sponsored by the Center for Urban Policy Research at Rutgers in New Jersey, or much of the material issued by the various trade groups in the building and housing industry. James W. Hughes, Rutgers, defines the 'excess costs' of government regulation to be those not 'necessary for protection of the basic health, safety and welfare of the occupants' (p. 318); the assumption is, not only that the authors know these necessities better than the legislature (which may be true), but that in the absence of regulations all housing would be built to the corresponding lowest standard (which is highly unlikely).

6 For an argument that home ownership differs only in minor and reversible detail from rental, but that social/ideological factors weigh heavily, see Marcuse, 1980.

7 See Engels, 1935; Angotti, 1977; and Achtenberg and Marcuse, 1986, pp. 474–83; and Brede, Kohaupt and Kujath, 1975.

8 See also Kastorff-Viehmann, 1914, and the work of Drs Franciska Bollery and Kristiana Hartmann. For the most famous case in the USA, see Buder, 1967.

9 See, for instance, Harloe, 1985.

10 Many factory owners shared Engels' views; the history of company housing reveals many employers who saw company housing as a direct method of controlling wage demands. Excellent documentation for such policies in the Ruhr is now available; see note 7 above. The idea is resurfacing in the USA.

11 Johnston, 1986, makes the argument explicit: 'Job markets were shifted, from housing markets containing "undesirable" workforces to those containing "desirable"' (p. 332). The empirical basis for the claim is somewhat weak; David Gordon, who first made the argument, has essentially only one quotation to substantiate it. Yet the argument seems logically sound. Further research is certainly warranted.

12 This is of course only part of an alternative explanation: the stimulation of consumption demand, for example, for household goods and for automobiles, racial prejudice, social status, are also part of the picture. For a fuller discussion, see Marcuse, 1988.

13 'Neighbourhood stability' as a watchword can mean very different things to different people/groups. For the poor, it can mean survival. For the insecure working class, it can mean protection of the only savings/security they have, in a home, against perceived threats

to investment or to status. For the upper middle class, it can mean resistance against the redistribution of environmental benefits downward, or a residential protest against business incursions. Class analysis of the forms of neighbourhood resistance to change is badly needed.

14 Johnston makes the same point somewhat differently; he refers to the priority of the National Union of Mineworkers in maintaining jobs *in place* as evidence (1986, p. 333).

15 The term is a good one, and deserves wider use. It was coined, I believe, by Paul Baran, to distinguish intellectuals – a term he reserved for those committed to a critical intelligence – from those who worked with their minds, rather than in routine physical or clerical or service occupations. Most intellectuals are intellect-workers, but not vice versa. See *Monthly Review*, May 1961.

16 Although not necessarily; see the recent Hollywood screenwriters' strike, and the additional independence generated by homework for multiple employers.

17 For a formal definition, see Marcuse, 1985.

18 Some of course are; if so, and they succeed, they leave the ranks of employees and become employers, capitalists in their own right, no longer working for others. In effect, they switch class. Most professional, technical, managerial employees don't.

19 The extent to which these attributes of workers' housing were matters of direct awareness and concern to employers can be seen through the history of company towns and employer-provided housing referred to above. The suburbs, *mutatis mutandis*, fit into the same tradition. See Gans, 1967.

20 For a good and critical overview of recent work, see Murie, 1986, pp. 345–57; Preteceille, 1987, for a solid theoretical discussion. Among the most interesting US empirical work is Halle, 1984, and Katznelson, 1981.

21 See the more fully developed discussions of production-based definitions in Poulantzas and Wright.

References

Achtenberg, E. and Marcuse, P. (1986), 'Toward the Decommodification of Housing', in Bratt *et al.* (eds), *Critical Perspectives on Housing* (Philadelphia, Temple University Press).

Angotti, T. (1977), 'The Housing Question: Engels and After', *Monthly Review*, October.

Brede, H., Kohaupt, B. and Kujath, H.-J. (1975), *Ökonomische und politische Determinanten der Wohnungsversorgung* (Frankfurt, Suhrkampf Verlag), no. 745.

Buder, S. (1967), *Pullman: An Experiment in Industrial Order and Community Planning, 1880–1930* (New York, Oxford University Press).

Burchell, R. and Listokin, D. (1980), 'The impact of local government regulations on housing costs and potential avenues for state meliorative measures', in G. Sternlieb, J. Hughes and Center for Urban Policy Research Staff, *America's Housing: Prospects and Problems* (New Jersey: Center for Urban Policy Research, Rutgers University).

Dolbeare, C. (1985), 'Federal housing assistance: who needs it? who gets it?' a policy working paper for the National League of Cities, Washington, DC.

Engels, F. (1935), 'The Housing Question', in *Handbook of Marxism* (New York, International Publishers).

Gans, H. (1967), *The Levittowners* (New York: Pantheon Press).

Gordon, D. (1984), 'Capitalist development and the history of American cities', in W. Tabb and L. Sawers (eds), *Marxism and the Metropolis*, 2nd edn (New York, Oxford University Press).

Halle, D. (1984), *America's Working Man: Work, Home and Politics Among Blue-Collar Property Owners* (Chicago, University of Chicago Press).

Hamnett, C. and Randolph, B. (1986), 'Socio-tenurial polarization in London: a longitudinal analysis' (Milton Keynes, The Open University), typescript.

Harloe, M. (1985) *Private Rented Housing in the United States and Europe* (London: Croom Helm).

Hoenderdos, W., Hooimeijer, P. and van Hoorn, F. (1986), 'Urban housing and labour markets in transition: research in progress', in *Tijdschrift voor Economic en Social Geografie* 77, No. 5.

Hughes, G. and McCormick, B. (1987), 'Housing markets, unemployment and labour market flexibility in the UK', *European Economic Review* 31.

Jackson, K. (1985), *The Crabgrass Frontier* (New York, Oxford University Press).

Johnston, R. J. (1986), 'Job markets and housing markets in the "developed world"', *Tijdschrift voor Economic en Social Geografie* 77, no. 5.

Kastorff-Viehmann, R. (1981), *Wohnungsbau für Arbeiter: Das Beispiel Ruhrgebiet bis 1914* (Aachen, Klenkes).

Katznelson, I. (1981), *City Trenches: Urban Politics and the Patterning of Class in the United States* (New York, Pantheon Press).

Kozol, J. (1988), *Rachel's Children* (New York, Pantheon Press).

Marcuse, P. (1980), 'The ideologies of ownership and property rights', in *Housing Form and Public Policy in the United States*, Columbia Monographs on Architecture, Preservation and Planning (Praeger Publishers).

Marcuse P. (1988a), 'Do cities have a future?' in *The URPE Crisis Reader* (New York: Monthly Review Press).

Marcuse P. (1988b), 'Preliminary analysis of NYCHA occupancy characteristics figures', unpublished.

Murie, A. (1986), 'Social differentiation in urban areas: housing or occupational class at work?' in *Tijdschrift voor Economic en Social Geografie* 77.

Poulantzas, N. (1968) *Political Power and Social Classes* (London, New Left Books).

Preteceille, E. (1987), 'Le concept de classe sociale dans la recherche urbaine', Centre Culturel Canadien, mimeo.

Rose, D. (1984), 'Rethinking gentrification: beyond the uneven development of marxist urban theory', in *Environment and Planning 2*, 47.

Sternlieb, G. and Hughes, J. (eds) (1976), *Post-industrial America: Metropolitan Decline and Inter-regional Job Shifts* (New Brunswick, Center for Urban Policy Research).

(1961), 'The Commitment of the Intellectual', *Monthly Review*, May.

Weisser, M. (1975), 'Arbeiterkolonien – uber die Motive zum Bau von Arbeitersiedlunger durch industrielle Unternehmen im 19. und fruhen 20. Jahrhundert in Deutschland', in Joachim Petsch, hrsg., *Architektur und Städtebau im 20. Jahrhundert*, vol. 2 (West Berlin, VSA), pp. 22, 23, 36.

Wirth, L. (1938), 'Urbanism as a Way of Life', *American Journal of Sociology* 44, pp. 1–24.

Wright, E. O. (1978), *Class, Crisis and the State* (London, New Left Books).

5 *The restructuring of work and home: productive and reproductive relations*

SOPHIE WATSON

In periods when major changes are taking place in the nature of the economy and social relations, it becomes imperative to re-examine concepts that we have inherited from the past. It is often only at such times that the value and usefulness of earlier forms of conceptualization become clear. Concepts that are used are appropriate to a particular point in the development of capitalism and need to be refined and re-examined as capital restructures.

This chapter is situated in the context of the realignment of economic social and political forces which have emerged during the 1980s. The concepts that it examines are work and home. In the last couple of decades the two spheres tended to be analysed separately, partly because there was less necessity to relate them, in that the dichotomy also reflected the dominant social relations of the time, and partly it was a reflection of 'objective' material conditions.

Many aspects of work did leave the home with capitalist development. However, it is the focus on this separation rather than on an analysis of shifting boundaries, and the consequent lack of recognition of other forms of paid work, notably home-based, which has contributed to the dichotomized conceptualization. The recognition by some feminist writers that the spheres were interrelated represents the major exception.

Feminist historians (for example, Davidoff, 1979; Matthews, 1983) have argued that traditional male categories of work have tended to obscure women's involvement in paid employment. This is derived from an emphasis on the more visible forms of work organized in 'collective' or public workplaces. Marxists' focus on the capitalist mode of production has been key here. Thus women's work, such as the provision of board and lodging or the private sale of skills, such as laundry, ironing, etc. to other households, was ignored both because of its location in the home and its location within the informal capitalist economy. These forms of labour are of increasing relevance in the current period as is explored in this chapter and challenge traditional notions of what constitutes 'work'. This is also a central feature of Pahl's (1984) thesis which recognizes in particular the absence of women's work – formal, informal or within the household – within standard analyses of work.

In hindsight it is clear that the home-work and public–private dichotomies reflect, in part, a real separation that occurred historically. The argument here, however, is that this separation, in so far as it did exist, is now breaking down. New interrelationships have formed and old dualisms are no longer useful. Production and reproduction are interdependent; neither independently determine the other. Changing forms of production and reproduction thus have to be viewed as part of a single and unified process.

This dualism has been central to Marxist definitions of class in that the separation of home and work was seen as central to the capitalist mode of production. Feminists' discomfort with this separation has been paralleled, therefore, by their discomfort with Marxist definitions of class. It is argued that a Marxist formulation of class has been phallocentric in that traditionally male forms of paid labour are taken as universal categories of work. The working class is seen to be of one gender (Pringle, 1989). This paper is part of a feminist challenge to previous notions of class and work. A focus on the interrelationships between home and work enables a new conceptual framework to be developed.

The chapter is divided into two parts. The first part addresses the concepts of home and work and argues that such notions are constituted by class, cultural and gender divisions. Once these dimensions are explored it is no longer possible to separate out home from work and complexities emerge. The dwelling, for example, may represent a 'home' for the man in a household and a workplace for the woman. For the purposes of analysis, home and work are initially treated separately in order to examine the distinctive nature of these concepts before exploring how they are interrelated. I then argue that notions of 'real work' have been constituted within patriarchal discourse by their spatial separation from the home. The second part of the chapter examines the interrelationships between home and work by looking at the links between housing and labour markets, at homework and tied accommodation, and at the new forms of the home/work interrelationship that are emerging. While the chapter is primarily conceptual it draws on empirical material from an Australian context. Rapid changes in the economy and urban form of the major cities over the last decade make Australia an interesting case study.

Home and work

Home

There is no straightforward definition of the 'home'. I have argued elsewhere (Watson, 1986a) that there is an important distinction to be drawn between a 'house' and a 'home'. The former is generally taken to be synonymous with a dwelling or a physical structure, whereas a 'home' is not. The 'home' implies particular social relations, or activities within a physical structure, and is strongly linked with a notion of family. Redclift (1984, p. 97) points out that the centrality of the production of use values and the reproduction of labour to conceptions of the domestic domain is somewhat tautological. 'Are the processes to be defined by the fact that they occur in a particular place, "the home", or is our idea of what constitutes the domestic defined as the site of certain specific processes?' she asks. Likewise notions of home are gender specific (Watson, 1986a), and also culturally specific. For a woman, the home may imply place of work, or a prison, or the site of violence, while for a man it may imply a place of leisure or somewhere he can exercise tyranny and control. In other words, the traditional sexual division of labour and gender ideology has implications for the concept of home. Further, the home represents a significant site of consumption and women's role as consumption agent within the home is an important one but this also varies between developed and less developed countries. As Redclift (1984, p. 108) points out 'in rural areas of developing countries where there are fewer commodities to "administer", the physical fabric of the dwelling, or the lack of it, allows for little fetishims. The domestication of women around commodity management is less pronounced'.

There are also class divisions in the meaning of the 'home' and like gender and cultural divisions, these permeate the interrelationship between home and work, and the level of self-provisioning within households as is discussed shortly.

Looking at the concept of the 'home' historically it is clear that the home has had, and continues to have, a quite different and shifting meaning for the working and middle classes, and for working- and middle-class women in particular. Regional disparities provide another dimension. Working-class households' dwellings have fewer and smaller rooms, less extra space, lower standards, poorer conditions and are far more likely to be rented than middle-class households' dwellings. What this implies is that there is less room for recreation, for visitors to stay, for the children to play on their own and for individual privacy, which involves differential levels of domestic work and child supervision for women. Further, rental tenure structures imply less freedom to change, improve, expand etc., the dwelling with again important implications involved for the meaning of the home for working-class tenant households.

The growth in domestic technology and recreation technology forms a new basis for the concept of home shifting yet further on class lines. Increasing use of new forms of technology (for example, microwave ovens) is apparent among higher income households. Contrary to expectation, research evidence suggests that this technology has not led to decreased time spent in different forms of domestic labour to evolve, with again implications for the meaning of home for middle- and working-class households. Likewise, higher income households have access to videos, computer games, etc., which enable recreation to take place within the home. Not only does this impact on public provision (for example, the closing down of local cinemas as they become increasingly unprofitable), it reduces working-class access to entertainment and hence exacerbates class divisions yet further. But also, households become increasingly privatized and isolated within the home. Toffler (1980) amongst others, has argued that the re-creation of a home-centred society is fundamentally positive. To the contrary it appears that there are widening class divisions in relation to the meaning of the home and increasing home centredness for different classes, which derive from differential access to employment and income.

Definitions of homelessness or lack of a home also shift culturally, historically and across class and gender lines. Thus, notions of security, control, privacy, emotional well-being, etc., are as constitutive of the notion of homelessness as are lack of a roof or poor material housing conditions. Yet definitions of the concept circumscribe the extent of the problem. So, single female homelessness remains hidden while the single homeless are defined as those living in hostels, since many of these restrict access to women (Watson, 1986a).

Work

Work is a similarly problematic concept and again gender, class and cultural divisions introduce further complexities. Part of the obscurity of the concept is constituted through the home/work relation. There is an intricate relation between women's domestic activity and their paid employment outside the home. Not only does female paid employment in the 'productive' sphere mirror the productive work they perform at home, but also women's work is defined as unskilled precisely because the skills required have been gained by women in the private domain. Thus, predominantly female work – such as catering, cleaning, sewing, etc. – can be performed with the minimum of training because a women's apprenticeship has taken place at no cost to the employer within the home. Pringle (1987) in her study of secretaries extends the argument suggesting that sexuality is crucial for an understanding of the workplace, while at the same time arguing secretarial work is not seen as 'real' work because it encompasses so many traditional domestic duties (consider the notion of the office wife). Thus, forms of work cannot be understood outside of the social relations within which they take place. Doing one's own or family's washing within the home constitute individual reproduction whereas taking in other workers' washing constitutes social reproduction. Likewise, making bread for self-consumption – which is a reproductive activity – differs from making bread to sell in the market-place – which is a form of commodity production. The complexity of the relations involved is evident.

Pahl (1984) differentiates the formal, informal and household economies within which work takes place. Pahl's work supports the argument that notions of work have hitherto been dominated by male paid employment. He draws attention to the fact that women have earned money for the entire history of wage labour and have contributed this income to their families and households (Pahl, 1984, p. 84). It is the emphasis on employment and the individual earner rather than the household as the basic economic unit which has led to the forms of work with which women have been primarily engaged becoming hidden. Feminist historical investigation of work provides similar insights. Matthews (1983, p. 13) suggests that the dimensions of time and space have served to obscure the view of women as workers. Men's work historically has been characterized by entering paid employment or a career outside the home and remaining within it until retirement or death. Women's employment, on the other hand, has been characterized by its proximity to, or location in, the home and is commonly interrupted by periods of childbirth and childrearing. Domestic service, for example, is frequently ignored in historical analysis of the economy, work and capitalism even though approximately one quarter of the female population of England, France and Australia were in domestic service at some point during their lives during the nineteenth and early twentieth centuries. Interestingly, the increase in homebased work and outwork for women with the current restructuring is similarly unrecorded and forgotten.

In the current period the concept of work is becoming increasingly problematic and is likely to become even more difficult to define as the labour process and labour force are restructured. This has always been in part the case with women's work, but men are also being increasingly drawn into patterns of work which involve part-time hours, periodic employment and unemployment, outwork, job-sharing and breaks in employment due to domestic responsibilities, notably childcare. Lack of work, or rather unemployment, is a particularly difficult concept to define, as is lack of home or homelessness. Moreover, definitions of female unemployment or underemployment pose specific problems, and consequently it is often unrecorded.

As noted earlier, female homelessness is concealed through lack of provision, forms of statistical analysis, and dominant notions of homelessness as male. Female unemployment is hidden also, through the structure of the social security system which is based on the notion of the male breadwinner and dependent wife, and again through the methods of collection of statistics and dominant notions of unemployment as male. In Australia annual Commonwealth Employment Service statistics on women's unemployment are underestimates for a range of reasons: the statistics do not cover people wanting to work less than thirty-five hours; women with employed husbands are unlikely to register because they do not qualify for unemployment benefit; women who are in receipt of social security benefits may not register as unemployed, even if they are seeking work because unemployment benefits are lower than the social security payments; and women who are tied to a particular locality because of family responsibilities or a husband's job, may not register if there are no job opportunities in that area (Women's Bureau, DEIR, 1985). Another problematic concept is underemployment, that is, underutilization of the time an employee is willing to work (involuntary part-time work) or underutilization of productive skills (when a person is working in a job at a lower skill level than their level of skills). Both these forms of underemployment are increasing at the present time. We can see from these examples how concepts of work are shifting and will continue to shift with restructuring.

Home and work – the interrelationships

Housing and labour markets

The interrelationship between housing and labour markets hinges on spatial and economic dimensions. Attention to gender divisions provides a challenge to traditional dichotomies. A family residence tends to be selected on the basis of its spatial relationship with the employment of the male partner, irrespective of female employment opportunities in the locality. Yet, women's primary domestic responsibilities restrict their ability to travel distances to paid employment due to lack of childcare services in particular. If the childcare costs are supported by women's wages, childcare costs interact with location in that the greater the amount of time a woman spends in travelling to work, the higher the childcare costs and the lower her net income from paid employment. Consequently women frequently compete for the relatively few jobs that are located within their area. But at the same time, married women enter the workforce to sustain mortgage repayments which are increasingly unviable for one wage earner to support in the major Australian capital cities where house prices are high (King, 1986, p. 280). Thus, there is a complex interaction between women's domestic burden, housing costs, location, the availability of childcare, lower female average earnings which influence the extent to which women are employed part-time involuntarily or are underemployed, and this further reinforces labour market segmentation on the basis of gender (Women's Bureau, DEIR, 1985). Poor public transport facilities represent another constraint, since many women do not drive or have secondary claim on the 'family car'.

Planning authorities recognize the connection but interestingly in many cases the response is a regressive one – to reinforce women's marginalization in the domestic sphere. Massey (1984, p. 230) quotes from the Cornwall Industrial Development Association plan: 'The problem as we see it, is not to bring in more female-intensive factory jobs; the problem is to take work to women to carry out at or near home.' The current NSW government investigation of the underuse of women's skills and employment potential in the western suburbs of Sydney is in danger of leading to a similar proposal.

Again, housing/labour market relations reflect the work/home dichotomy and dominant gender relations. Yet access to housing and access to employment can be seen as part of a unified picture. Income is necessary to buy housing, where housing is primarily a commodity produced for a profit. But housing location, in particular, constrains access to the paid employment where income can be generated – particularly where women are concerned. The locationally fixed nature of housing is also important in that in areas where manufacturing industries are declining or moving elsewhere, households become trapped in housing that is difficult to sell. Sometimes these are precisely the same (or close to) areas, where new sectors of employment are expanding, and the resident households are hampered by the lack of necessary skills.

In inner Sydney between 1976 and 1981 16,000 jobs were lost in the manufacturing sector, at the same time as Sydney's growth as an international financial centre led to an increase of 40,000 jobs in finance property and business services. What this has led to is greater concentration of unemployment in the poorer inner areas. Contrary to the expectation that married women whose husbands were retrenched would enter the labour force to bring income into the household, in areas where there is an increase in unemployment, married women's labour force participation has declined. The converse pattern also holds. Thus, wives of unemployed men are six times more likely to be unemployed and half as likely to be in paid employment compared with wives of employed men (Bradbury *et al.*, 1985, p. 9). This can be explained by the effects of economic restructuring, and a mismatch between household characteristics and a new demand for labour in the finance and business services sector to which retrenched manufacturing workers have no access, either for location reasons or because they lack recognized skills. Between 1981 and 1986 there was an increase in the central Sydney area of professional and para-professional workers from 15 per cent of the region's labour force to just over 20 per cent (Searle, 1987, p. 5). The operation of the benefits system provides a further explanation, since it acts as a disincentive to labour force participation amongst women married to unemployed men.

Home-work and tied accommodation

In the previous section some of the interrelationships between housing and labour markets have been established. So far housing and labour markets within this division have been treated as spatially distinct. Home-work and tied accommodation are interesting because they illustrate forms of interconnection between work and home where these co-exist in a single location. It will be argued that in both these situations relations of work and home, and class and gender are intertwined thus revealing further the falsity of the separation and the importance of establishing interrelationships.

The use of home-work by capital as a means of reducing the costs of production (no overheads, workers frequently pay the cost of the means of production, etc.) and combating forms of collective organization within the workplace is not new. However, in the current period of economic restructuring, and with the emergence of a new international division of labour, a shift from factory production to home-based production appears to be an important part of the restructuring process (Harrison, 1983; Huws, 1984; Mattera, 1985) and is once again attractive to capital. New technology has played an important role as industries facing declining profitability and increased competition off-shore (particularly from South-east Asia) have reorganized employment as a means of cutting costs. In the manufacturing industry, it has enabled many production processes to be simplified, so that 'skilled' work which previously took place in a factory can be performed by 'unskilled' workers at home. In Australia, 13,900 women and men were employed at home in 1981 as metal processing and electrical workers, food and tobacco processing workers, wood processing workers and as chemical and paper processing workers (Walker, 1987, p. 24). In the service sector, likewise, new forms of technology have been introduced which enable a diversity of clerical, computing and professional work to be performed at home. Moreover, new computerized equipment can monitor the user's speed, efficiency and working hours thus introducing direct outside managerial control (Bisset and Huws, 1983, p. 8).

Although home-work has traditionally been associated with female employment, there are increasing numbers of male home-workers also. In 1981 in Australia 221,100 men and 230,200 women were engaged in home-based employment. The almost equal numbers mask the fact that home-work constitutes a larger proportion of total female (9.8 per cent) than male (5.6 per cent) employment (Walker, 1987, p. 20). However, home-work has different implications for the two sexes. At the most basic level what home-work does is fundamentally change the meaning of the 'home' for the workers involved and challenges the traditional separation of the home and (paid) work spheres. The home becomes the workplace also. For the woman worker not only does this imply increased domestic work, pressure on space within the home, potential family tensions, but also women home-workers face double exploitation within the domestic domain from their employers as wage labourers and from their husbands as domestic labourers (Walker, 1987, p. 7). Further, where women are part of an extended family, where all the members are home-workers, as is particularly common amongst ethnic minorities, they are unable to establish a 'spatially separate sphere of existence' and 'within the family itself the double subordination of women is fixed through the mixing in one person of the role of husband or father with that of boss and employer' (McDowell and Massey, 1984, p. 145). Thus, home-work for women is likely to enmesh them further within a complex web of oppressive and exploitative domestic and labour relationships in a single location. Working for a wage at home enables women to continue to carry out domestic responsibilities, particularly childcare (indeed this is the reason many women engage in outwork), and the domestic division of labour is not threatened. Participation in the labour market is so marginal that women do not gain the autonomy or power that collective organization in the workplace and economic independence can imply, to challenge gender relations at home or the wage relation of the workplace. Wajcman and Probert's (1988) study of new technology 'Bringing it all back home: a study of new technology outwork' substantiated this argument. They found that men who worked at home as programmers came closer to the Toffler vision of a more flexible lifestyle, but that this did not contribute to a less rigid sexual division of labour within the home. Instead, men who worked at home became more work oriented and paid less attention to their families, whereas women became more family oriented.

The form of home-work varies considerably between classes. First, home-work does not necessarily imply low wages, poor conditions, insecure employment – although it often does. Some professional employment performed at home could be defined as home-work and with advances in new technology, and computerization a variety of skilled and professional jobs (for example, editing of manuscripts) could become home based. In Australia 33,100, or 7.3 per cent of all home-based workers are employed in the growth industries of financial, property and business services. The interesting dimension here is that gender divisions cut across class divisions. Thus there is under-representation of women in the professional sections of these industries such as architecture, surveying, accountancy and computer programming – the more 'skilled' areas and an over-representation of women in clerical jobs. This sexual division is starkly reflected in the disparity of incomes. In 1981, of the full time professional and managerial workers (comprising 60 per cent of the male and 22 per cent of the female home based workers in the industry), over 80 per cent had an income of over $10,000, while the comparable figure for clerical workers was only 43 per cent (Walker, 1987, pp. 38–40).

Second, there is an important relationship between tenure and/or housing standards and home-work. Traditional middle-class women's home-work, for example, piano teaching, personal dressmaking, etc. require a separate and private room, good housing conditions and even a favourable location. The structure of home ownership provision may mean that it is often the only form of tenure to have these attributes. Similarly, even the more working-class forms of home-work may require considerable space to locate the necessary machinery or to mind children, for example, and for manufacturing jobs particularly it may be necessary to convert garages into workshops or to construct extra rooms or extensions. Again, home ownership represents a less restrictive form of housing in this respect while public tenants are likely to be more constrained.

The interesting aspect of tied accommodation is that employer–employee and landlord–tenant relations coincide. A doubly exploitative relation becomes possible. Thus retrenchment or sacking from employment usually implies eviction also, and the two forms of security offered by housing and employment are interdependent. Women, who are employed in parts of the service sector which are low paid, are particularly locked in by their interrelationship. Two notable areas here are the hotel and catering trade and nursing where women predominate in low-paid jobs. In Australia the growth of the tourist industry is contributing to a significant growth in the number of jobs with accommodation attached. In 1974, 96,800 women and 76,000 men were employed in restaurants, clubs and hotels. By 1984 the numbers of women and men employed in the industry had increased to 98,800 and 119,800 respectively (ABS, 1985).

Low incomes constrain access to private sector housing (either rented or bought), and the ability to move jobs and to find independent accommodation is severely restricted. Thus, for example, a hospital nurse who wants to work in the community sector, or a hotel waitress who wants to work in an office canteen, not only has to obtain the alternative employment, but also has to find alternative accommodation before shifting jobs. The low income and further deductions for accommodation of the previous employment is liable to militate against such mobility. Further, many women who have no recognized skills, who are divorced or widowed in the latter part of their adult life, who have lost accommodation, have no option but to enter employment with accommodation offered as a combined solution to their housing and employment problems (Watson, 1986b).

New forms of the home/work interrelationship

The movement of reproductive activities in and out of the domestic sphere is constantly in a state of flux, and more specifically, these processes are changing once again with the current economic restructuring. The following arguments are speculative but will be raised nevertheless since they illustrate how new forms of interrelationships between the spheres of work and home are developing. To take Sydney as one locality, the current trends in female employment patterns suggest that the decline in the manufacturing industry combined with rationalization and decentralization to outer suburban areas, and the growth in the service sector, in particular, the growth of the banking, financial, insurance and property services, has important implications for home/work relations. In summary, married women's labour force participation is increasing in middle-class areas and decreasing in working-class areas (Tait, 1985). This reflects the growth in female 'skilled'/qualified as opposed to 'unskilled'/unqualified employment and is thus of no surprise. At the same time as working-class women's jobs are declining overall, there is a significant amount of recorded and unrecorded informal forms of work and outwork within this class. The extent to which this form of work is increasing is difficult to assess. However, the current shifts in Australia towards increased flexibility by the introduction of alternative working practices and de-regulation of the labour market is expected to include an expansion of home-based working (Walker, 1987, p. 3). Of relevance is the apparent increase once again in the commodification of domestic (reproductive) activities in certain areas. Childcare, laundry and domestic services have increasingly re-entered the sphere of paid work production as married women have entered the labour force (leaving aside for the moment the significance of married women's part-time employment and double work burden). However, these are not universal processes and what is interesting is that class divisions appear to cut across sexual divisions. That is, as reproduction is commodified it moves out of middle-class women's homes into working-class women's homes – with implications for class divisions in the home/work dichotomy – and into the sphere of capitalist production.

To take childcare, laundry and cleaning as three examples. Married middle-class women's labour force participation, particularly full-time, is only possible if large parts of necessary domestic labour are farmed elsewhere. With decreasing employment opportunities in the formal sectors, working-class women are performing this service. This occurs in two ways; one, they take middle-class families children into their home – childminding as it is known in Britain, family day care in Australia – which more correctly should be termed family outwork. There has been an increase in Australia of approximately 21,000 family day care places between 1976 and 1985, to approximately 30,000 places in 1985 (Office of Childcare, 1986), which corresponds with rising married women's labour force participation. Some of this work is formally recorded, while apparently similar proportions are performed on an informal basis, often by one working-class woman in her home for two or three professional women who are in paid employment. These women's incomes are high enough to support payment for the necessary aspects of their own and their household's reproduction. Within this scenario, the working-class woman is involved in two forms of reproduction – individual reproduction (such work is often combined with looking after her own children) and social reproduction. In this case middle-class women's childcare is thus commodified, while working-class women's is not. The other way paid childcare is undertaken, involves similar social relations to performing another household's housework. Here, one woman leaves her home and enters another woman's home to work – with different implications for the meaning of the home for both women involved. This form of paid employment is clearly part of the informal sector, and rarely recorded.

Exploitation occurs in a privatized manner and usually involves one woman employing another. Finally, other aspects of domestic labour amongst higher income women may be increasingly commodified in the public and regulated sphere. This is likely to occur with domestic work – such as laundry (i.e. service washes) which involved drudgery – freeing women (who have the choice) to engage in more 'rewarding' forms of 'self-provisioning' (for example, gardening). As Pahl (1984) demonstrates, higher income couples who are both employed full-time use more formal provision. If this became a growing trend, this would have implications for patterns of consumption within households. At the same time, while some domestic processes are being commodified, other forms of work that were previously centralized in the public sphere are being relocated in the household with the growth of new forms of technology such as the proposal for remote shopping (Gershuny, 1984).

The relationship between reproduction and household and demographic structures adds further complexity. As Mingione (1984, p. 28) argues, 'Industrial development progressively dissolves large households, based on extended families, in favour of much smaller ones, based on nuclear families . . . [with] consequences for the organization of reproduction cycles. Changing demographic patterns are thus important.

Throughout the industrialized world the population is ageing and marriage and birth rates are decreasing, young people are leaving home earlier to establish independent households and marital breakdown is increasing. Overall, this has led to an increase in one person households and single parent families. As Mingione (1984) argues, this influences reproduction cycles because these households generally have below average incomes and have to engage in informal, illegal and reciprocal consumption or production activities. Moreover, services provided by the welfare state are necessary to help reproduce these households and to prevent social conflict and to shore-up the consumption activity of these households. These demographic changes are of increasing relevance to the notion of the informal sector and self-provisioning.

The current interest in self-provisioning and informal work

reflects the recognition that old Marxist categories are no longer useful and a new conceptual framework is needed to analyse the changes in employment and work practices that are occurring as part of the restructuring process. It is argued that reduced formal employment and shorter periods of participation in the paid labour force (through early retirement etc.) tends to shift the balance between work for income and work for self-consumption in favour of the latter (Mingione, 1984). That is, there is a shift from the productive to the reproductive system. However, the restructuring of capital is very uneven in its effects in that households vary in their ability to improve their material position through self-provisioning and informal employment. In the final part of this chapter it is shown that divisions of gender, class and race are important here, as is household form and tenure.

Pahl's (1984) work is of particular interest in this context, since his focus is on the household rather than on formal labour markets. Through a detailed study of the divisions of labour between and within households in the Isle of Sheppey in Kent, he analyses the relationship between the forms of work undertaken by households (self-provisioning, informal work and paid employment) and the sources of labour households employ (formal provision, informal work and the household itself). In relation to class divisions, the study shows that the higher the household income and the greater the number of household members in the paid formal economy, the more likely it is that the household will engage in self-provisioning and use formal service provision (Pahl, 1984, p. 222). In contrast, the poorer the household and the lower the number of income earners within the household, the less the self-provisioning and the greater the reliance on informal sources of unpaid labour. This is an important finding since it contradicts the commonly held assumption that low-income or unemployed households compensate for their lower economic status by using their 'free' time to provide for themselves in the domestic sphere.

Analysis of demographic and household structures sharpens the divisions: there is a much lower level of self-provisioning within single parent households than in households comprising two parents. Instead, single parents and old people are obliged to use informal sources of labour due to lack of income (Pahl, 1984, p. 227). There is another issue here, however, which Pahl does not address, which is that research evidence in Canberra and Sydney suggests that women, particularly older women, are fearful of having trades*men* (my emphasis) whom they do not know into their dwellings through fear of harassment or being 'ripped off'. This is particularly the case for women who have previously relied on male partners to maintain, or arrange the maintenance, of the dwelling (Coleman and Watson, 1985, p. 4).

Nevertheless, levels of self-provisioning and use of formal and informal services obviously vary according to household structure and income. They also vary according to stage in the domestic cycle. Clearly families during childbearing and childrearing years spend more time in childcare, while households without children have more options to generate income through paid employment and through domestic commodity production, and to generate resources within the household through self-provisioning. Overall, Pahl's data can be explained at least partly by the higher levels of home, land, car, equipment, tool, etc., ownership among higher income groups. Lower income households which do not have access to ownership of these resources and commodities are hence less able to engage in self-provisioning activities.

In Australia, the lowest labour force participation rates are found in female single parent households (ABS, 1981). Female single parents are considerably less likely to be employed than their male counterparts and also married women with children (28.5 per cent of female single parents with the youngest dependent child under four years compared to 40.2 per cent of married women in the same category are in the labour force (ABS, 1984, p. 43)). Further, female single parents in Australia are considerably more likely to be public (25 per cent) and private (40 per cent) tenants rather than home owners/purchasers (27 per cent) while male single parents or male heads of nuclear families in the same age group are far more likely to be home owners or purchasers – 40 per cent and 55 per cent respectively (Watson, 1986b, p. 6). Thus, female single parents are not likely to engage in self-provisioning on income grounds alone but also on housing grounds, since lower home ownership rates and restrictions imposed by public and private landlords are likely to militate against such activities. (Landlord responsibility to maintain dwellings is also an issue here.)

Moreover, in the context of the spatial distribution of housing tenure and households in Sydney, there is a mutually reinforcing relationship between public sector housing and lack of access to the labour market. Public housing is predominantly located in the far western suburbs of the city, far from centres of employment and poorly serviced by public transport. These are the new release urban development areas where land is cheaper. There is thus a concentration of single parents, single older people and lower income young families in public housing in these locations. These are also households who have lower rates of private car ownership. The lack of access to home or car ownership implies exclusion from access to the labour market and vice versa. Thus these lower-income households, many of whom are women, become increasingly isolated and less likely to 'self-provide' since they cannot break the circle of marginalization.

The forms of this spatial separation of housing and labour markets or home and work are also undergoing change. The Sydney urban form was developed in the postwar period when women were primarily consigned to the home and a domestic role in the suburbs, which were removed from the centres of employment to which men went to work. Thus, the ideology of suburban home ownership living went hand in hand with a growing economy. It was only viable when the members of suburban households had jobs, could afford cars and could take on and sustain mortgage repayments. In the current situation, many households are trapped in the suburban home, unemployed, unable to afford a car, and certainly not two, and burdened with housing costs that they cannot afford. The inadequate public transport network itself reflects the fact that suburban development was based on the assumption of car ownership. The growth sectors of the economy are primarily located in the central business district of the city. Thus, the suburban home is shifting from being a retreat from the 'hurly-burly' of work (for men) to a prison for the entire household. Again we see the interrelationship between work and home undergoing a process of change.

There is a further aspect of self-provisioning and that is that female single parents face time constraints due to domestic responsibilities and are thus not frequently able to capitalize on existing resources. Moreover, these households and older women (the majority of the aged) are bound by culturally specific ideologies of what constitutes gender appropriate activities – hence the low levels of car maintenance (which are greater than would be expected from car ownership levels performed by single parent households in Pahl's study (1984, p. 227)).

Despite the useful contribution to the new forms of analysis advocated here, there are several problems with Pahl's study. First, the notion that households constitute a unified and cohesive structure. Households, particularly two parent households, breakdown with increasing regularity – one in three marriages in Australia ends in divorce. Thus levels of self-provisioning among couples can change quite dramatically. The conceptual framework therefore needs to be adapted to incorporate the notion of change and impermanence. The ideology of household unity which is dominant in Pahl's approach may conceal a quite diverse set of interests within the household. Edwards' (1981) survey of couples' finances found a great diversity of financial arrangements within households in Australia although women overall were less likely than men to control family finances, or to share in that control, whether or not they were in paid employment (p. 133). Thus even engagement in the productive sphere does not necessarily imply that women will gain control or power in the reproductive sphere. Hence, the benefits gained from paid employment, informal work, or self-provisioning are likely to be unevenly distributed within the household.

The second point is that the informal sector is sharply divided on the basis of gender. Women's lack of recognized marketable skills means that the informal sector for women is characterized by low pay for work done, home-based work and the decommodification of welfare services. The economic recession, the restructuring and rolling back of the welfare state in many capitalist economies and decreased public expenditure, has meant that women increasingly bear the burden of caring for dependants (young, disabled, aged, mentally infirm, etc.) within the home. In this context, the informal sector becomes a substitute for government intervention and expenditure at women's expense. In contrast, men within the household are more likely to have skills which can be exchanged in the marketplace and outside of the domestic sphere for an income which is untaxed and hence possibly higher than incomes for equivalent work in the formal sector. Once again gender divisions cut across household structures and the meaning of work. Moreover, there is a spatial dimension to the informal sector. Access to informal work varies according to the employment structure of areas and the relative strength of specific groups – women, immigrants, skilled workers etc. – in a given labour market. Thus, in areas where unemployed single parent households predominate and formal paid employment opportunities are scarce, there is likely to be lower levels of self-provisioning and a great reliance on informal labour, than in an area where there is a concentration of middle-income couple households and well paid employment opportunities.

Conclusion

In this chapter I have argued that dichotomies of home and work, production and reproduction derive from a Marxist framework which has focused on some forms of work at the expense of others. Thus home-work and other work practices located in the domestic sphere have traditionally been ignored. The current period of economic restructuring, where increasing numbers of people are engaged in a range of diverse and interrupted work patterns has coincided with the emergence of new conceptual frameworks which have been influenced by feminist analysis. The focus on an analysis of the shifting interrelationships between home and work offers the possibility of an account which challenges old dualisms and explores new definitions of class from a feminist perspective. It is to be hoped that this chapter contributes to such a project.

References

Austerberry, H. and Watson, S. (1981), 'A woman's place. A feminist approach to housing in Britain', *Feminist Review*, 8.

Australian Bureau of Statistics (1981), One per cent household file.

Australian Bureau of Statistics (1984), Australian Families 1982, (Canberra, ABS).

Australian Bureau of Statistics (1985), 'Labour Force Australia Historical Summary', Cat. 6204.0 (Canberra, ABS).

Bisset, L. and Huws, V. (1983), *Sweated Labour* (London, Low Pay Unit).

Bradbury B. *et al.* (1985), 'Bearing the burden of unemployment – unequally: a study of Australian households in 1981', Social Welfare Research Centre Proceedings, 53 (Sydney, University of New South Wales).

Bruegel, I. (1979), 'Women as a reserve army of labour: a note on recent British experience', *Feminist Review*, 3, pp. 12–23.

Coleman, L. and Watson, S. (1985), 'Older women: housing issues and perspectives', *Australian Journal on Ageing*, vol. 4, no. 3.

Davidoff, L. (1979), 'The separation of home and work? Landladies and lodgers in nineteenth and twentieth century England', in S. Burman (ed.) *Fit Work for Women* (London, Croom Helm).

Gershuny, J. (1984), 'Economic development and change in the mode of provision of services', in N. Redclift and E. Mingione (eds), *Beyond Employment: Household Gender and Subsistence* (Oxford, Basil Blackwell).

Harrison, P. (1983), *Inside the Inner City* (Harmondsworth, Penguin).

Huws, U. (1984), *The New Homeworkers* (London, Low Pay Unit).

King, R. (1986), 'Housing policy, planning practice,' in B. McLoughlin and M. Huxley, (eds), *Urban Planning in Australia: Critical Readings* (Melbourne, Longman Cheshire).

McDowell, L. and Massey, D. (1984), 'A woman's place?' in D. Massey and J. Allen (eds), *Geography Matters* (Cambridge, Cambridge University Press).

Massey, D. (1982), *Anatomy of Job Loss* (London, Methuen).

Massey, D. (1984), *Spatial Divisions of Labour* (London, Macmillan).
Mattera, P. (1985), *Off the Books* (London, Pluto).
Matthews, J. (1983), 'The Proletarian's Wife', *Politics*, vol. 18, no. 2, pp. 104–7.
Mingione, E. (1984), 'Social reproduction of the surplus labour force: the case of southern Italy', in N. Redclift and E. Mingione (eds), *Beyond Employment: Household, Gender and Subsistence* (Oxford, Basil Blackwell).

Pahl, R. (1984), *Divisions of Labour* (Oxford, Blackwell).
Pringle, R. (1987), 'What is a secretary?' Paper presented to the third International Women's Studies Conference, Dublin.
Pringle, R. (1989), *Sexuality, Power and Work: Secretaries* (Sydney, Allen & Unwin).

Redclift, N. (1984), 'The contested domain: gender, accumulation and the labour process', in N. Redclift and E. Mingione (eds), *Beyond Employment: Household, Gender and Subsistence* (Oxford, Basil Blackwell).
Redclift, N. and Mingione, E. (eds) (1984), *Beyond Employment: Household, Gender and Subsistence* (Oxford, Basil Blackwell).

Searle, G. (1987), 'Getting to work in Sydney: spatial balance between labour and jobs now and in future'. Paper presented to the Annual Conference of the Regional Science Association Australia and New Zealand Section, Deakin University, 9 December.

Tait, D. (1985), 'Jobs and poverty: the changing face of Sydney', Working Paper, SWRC (New South Wales University).
Toffler, A. (1980), *The Third Wave* (New York, William Morrow).

Vanek, K. (1974), 'Time Spent on Housework', *Scientific American* 23 November, pp. 115–20.

Wajcman, J. and Probert, B. (1988), 'Bringing it all back home: a study of new technology outwork', in E. Willis (ed.) *Technology and the Labour Process: Australian Case Studies* (Sydney, Allen & Unwin).
Walker, J. (1987), 'Home-based working in Australia: issues and evidence', URU Working Paper no. 1 (Canberra, Australian National University).
Watson, S. (1986a) *Housing and Homelessness: A Feminist Perspective* (London, Routledge & Kegan Paul).
Watson, S. (1986b) 'Women and housing or feminist housing analysis?', *Housing Studies*, vol. 1, no. 1, January.
Women's Bureau, DEIR (1985), 'Women and Unemployment', Information Paper no. 2 (Canberra, Department of Employment and Industrial Relations, Women's Bureau).

PART THREE

Mapping the connections

Introduction

The previous section examined a number of different ways in which the analysis of labour and housing market interactions can be conducted. In this section the focus of attention shifts to look at the variety of interactions across the UK. This is an issue of considerable importance for two reasons. First, as we stressed in the general introduction, and Randolph argued in Chapter 1, labour and housing markets have their own distinctive geographies. The structure of employment opportunities at any one point in time is spatially fixed, as is the occupational structure of the paid labour force. And, to state the obvious, housing is a spatially fixed commodity par excellence. Once built, houses cannot be moved.

But while houses and places of work are fixed in space, labour is mobile and homes and work places are linked through journey to work trips. It is the inherent spatiality of both employment and housing, and the friction of distance imposed by journey-to-work times and costs that link labour and housing markets in space. Were this not so there would be no spatial linkage between place of work and place of residence.

As a result of the inherent spatiality of the two markets, the structure of the two markets and the interaction between them are highly variable over space. And, because the structure of either market at any point in time reflects and embodies their historical development, there is also an inherent temporal variation in the two markets and their interactions. In the nineteenth century and earlier, when transport was poor and travel to work distances were very limited, labour and housing markets were more tightly bound together than they are today (Vance, 1966; Steadman-Jones, 1971).

While the greater mobility of the postwar period has weakened the friction of distance and the locational constraints between homes and workplaces, and has allowed greater journey-to-work distances for some social groups, constraints on mobility are still marked. For most people, labour market opportunities are limited to what is available within easy or affordable daily commuting distances. And the structure of both labour and housing market opportunities in specific areas is given by the historical development of those areas. A legacy of de-industrialization, high unemployment and a preponderance of council housing or low priced nineteenth-century terraced housing poses a very different set of opportunities and interrelationships from areas with a growing service sector with large numbers of professional and managerial jobs, low levels of unemployment and a high priced owner-occupied housing market.

As a result of these spatial and temporal variations, there are very marked differences in the structure of individual labour and housing markets and their interrelationships. This implies, as Coombes, Champion and Munro point out in Chapter 6, that a focus on local labour/housing market interactions is important for understanding the formal, as opposed to the purely conceptual, links between the two spheres on the ground. However, our interest in geographical variation is not simply that the legacy, structure, and interrelationships between the two markets differ over space. We are not interested in differences just for their own sake, important though they may be.

Such differences play an active part in shaping the evolution of labour and housing markets in Britain. They have an impact on both the local variations themselves and on the overall structure of housing and labour markets in Britain and in other countries. The differential reshaping of local labour and housing markets affects the structure of the whole economy. De-industrialization in some areas and the emergence of service-based economy in others, has produced a new geography of labour market opportunities while the differential pattern of housing tenures and house prices has produced a new geography of housing opportunities and constraints. This is the second reason for focusing on geographical variations.

Although the chapters in the section look at housing and labour markets in a variety of different areas of Britain, a number of common themes unite them. The first such theme is the extent to which labour and housing market changes compound or reinforce one another and lead to advantaged and disadvantaged areas. This can happen at both inter- and intra-regional scales. At the intra-regional scale it has been argued that many inner-city residents have been trapped by the housing market and locked into declining economies (Metcalf and Richardson, 1980 and Hamnett and Randolph, 1986). And at the inter-regional scale it has been argued that labour migration in Britain has been limited by both council housing (Hughes and McCormick, 1981, 1987; Minford *et al.*, 1987) and by marked regional variations in house prices. This division has been well documented (Hamnett, 1988, 1989a, b).

The argument that differences in regional house prices and rates of accumulation reinforce the inequalities arising from regional variations in labour market conditions and industrial structure was made by Thorns (1982). He suggested that the older industrial regions of the Midlands and the North were characterized *both* by high levels of job losses and unemployment compared to the more prosperous areas of the South-east, *and* by a depressed owner-occupied market in which both prices and rates of house price inflation were much lower than in the South-east. He argued that a process of cumulative and progressive regional disadvantage was operating. This disadvantage was opening up a growing gap between the affluent and the depressed regions, which was simultaneously providing both jobs and equity accumulation in the south as well as locking residents of depressed regions into declining labour and housing markets.

But, valuable though Thorns' work was in pointing to the links between labour market restructuring and the property market in general terms, he did not undertake any systematic empirical analysis of regional variations in house prices or rates of house price inflation. Nor did he attempt to systematically link them to labour market conditions (see Hamnett, 1984, 1989 for a critical appraisal). The value of the chapter by Coombes, Champion and Munro lies precisely in its attempt to link house price variations to measures of labour market conditions in local labour market areas across Britain as a whole. Their starting point is that contrasts in local housing demand pressure flow from variations in local economic growth or decline. They then seek to 'investigate the causal processes underlying particular levels of prices, or rates of inflation, by placing the local house price "outcome" in the context of demand and supply factors within each housing market' (p. 173). Their analysis is based on house price data and several labour market indicators across 280 'local labour areas' in Britain, and variations in house prices are taken as the dependent variable which is to be statistically explained by regression analysis of the economic/labour market variables.

They argue that this approach is important in so far as previous studies have given little guidance as to what are 'the important labour market factors to include in an empirical investigation of local housing outcomes.' And they suggest that the specification of the model has 'to be both conceptually and empirically driven, as it is possible to postulate a wide variety of linkages, none of which are obviously superior to all other alternatives' (p. 180). To this end, they first correlate house prices with the 'index of local economic performance' which includes three indicators of population and economic change, the unemployment rate and car ownership as a surrogate of affluence and income. This produces a high level of correlation and they then proceed to disaggregate the index using regression analysis. They show that the variable which is consistently most strongly correlated with house prices is the proportion of the workforce who were in professional or managerial occupations, followed by unemployment, employment change 1971–81, car ownership, the proportion of owner-occupied households and the proportion of workers employed in finance. The problem with this sort of analysis, of course, as the authors recognize, is that the causal nature of the links still requires further theoretical analysis.

The nature of some of these causal links is taken up in the other chapters in this section in a variety of ways. One of the most important of these, which forms the second theme, is the position of different areas within the national and international spatial division of labour, their different industrial histories and occupational structures, and the impact of changes in their position on labour and housing markets. This is clearly marked in terms of the division between old declining industrial cities and regions, and growing sunbelt cities and regions with a high-tech and services employment base.

The chapters cover a wide range of locations in this respect. At one end of the spectrum, Hamnett (Chapter 7) examines the role of London as a global headquarters of multinational capital and finance, and looks at the extent to which intensification of this role is associated with a changing industrial and occupational structure, and the implications of this for the housing market. Barlow and Savage (Chapter 9) look at the related case of Berkshire which was previously a middle-class commuting area for London, but which is now one of the growth centres of sunbelt Britain and a centre of high-tech industry, R and D and financial services. They show how the region's character has changed as manual jobs in manufacturing have disappeared to be partly replaced by professional/managerial jobs in the new growth sectors. And Goodwin (Chapter 10) looks at the rapid transformation of London's Docklands from a rundown port and industrial area to a key centre of financial and allied services and luxury housing which is closely linked to London's role in the global financial economy.

At the other end of the spectrum, Hudson (Chapter 8) examines labour and housing market links in a peripheral region – the North-east. He shows how the economic role of the North-east has been changed by the post-1974 recession, firm closures and unemployment from that of a major coal mining and heavy industrial area to a cheap labour reserve and branch–plant economy, first for national, and more recently for multinational capital. As a consequence, the economic role of the North-east has shifted from a centre of heavy industry, first to a 'national' and then a 'global' assembly outpost.

And the composition of the labour force has shifted away from skilled male labour towards semi-skilled female labour.

These changes in the position of different areas in the international economy are related to uneven development and the impact of changes in the pattern of capital investment and disinvestment. Goodwin argues that changes in housing and employment markets have played a key role in the process of uneven development in Britain. He argues that uneven development is inevitable where capital is constantly moving from one area and region to another in search of greater profitability and accumulation of surplus value. While new capital is invested in some areas, capital is continually withdrawn or devalorized in others. The result is that social and economic geographies are continually being developed, changed and abandoned. Different areas thus represent different stages in this restless search for profitability. While some areas may be experiencing a boom in investment and, possibly, in employment, other areas may be experiencing widespread industrial closures and the decline or abandonment of traditional industrial bases and their replacement by another type of economic activity, with very different labour demands and requirements.

Goodwin suggests that different periods and forms of production and consumption are physically embodied through 'the creation of fixed and relatively immobile spatial configurations' which have an individual 'structured coherence'. In other words, different structures of housing and labour markets become fixed in space. But, as the requirements of capital change, these geographical concentrations of investment, labour and production may become redundant, and they are replaced by other assemblages as one 'spatial fix' is replaced by another.

This point comes across very clearly in the case of London's Docklands, where Goodwin shows how the traditional port-related processing industries have been replaced by new financial services with very different labour demands. As a result, the old skills of the area have been rendered redundant; they are not appropriate to the new labour market which is being created by the London Docklands Development Corporation (LDDC). The traditional workforce and housing market structure has become surplus to the requirements of capital, and is in the process of being replaced by new, more 'appropriate' structures suitable for late twentieth-century forms of capital accumulation.

Hudson is also able to show the operation of similar processes in the North-east. Here the collapse of the traditional mining and heavy industries have rendered large parts of the existing male labour force redundant, and led to the growth of branch plants employing 'green' female labour. Hudson argues that this has had major implications for the housing market: 'the precipitate decline in labour demand has rendered parts of the housing stock as redundant as the workers who once lived in it.' More recently, the existence of a tradition of male industrial labour has attracted companies such as Nissan into the region in search of a European assembly base. Barlow and Savage show that in Berkshire traditional manufacturing industries employing skilled and semi-skilled labour have also been partly replaced by high-tech or financial services employment with very different labour demands for higher skilled professional and technical workers.

The differential adjustment between housing and labour markets is, therefore, an important focus of this section. But this is not to say that the process of adjustment to economic growth and decline is similar to the instantaneous equilibrium of neo-classical economics. Because housing is immobile and because people are more immobile than jobs, local residents are often locked into areas of economic decline. The reverse is true in the growth areas where employment is growing, and where home ownership is dominant and house prices are high. In such areas, despite high levels of new building, there may well be severe housing shortages, particularly of affordable housing. Such shortages affect the less skilled and less well paid more than most, and they are often marginalized in the housing market (Bramley and Paice, 1988).

The replacement or displacement of different sections of an area's workforce or resident population as a result of changes in labour and housing markets (and associated economic and social polarization) is the third main thread running through the chapters. Hamnett argues in Chapter 7 that shifts in the industrial structure of London, as manufacturing industry has declined to be partly replaced by financial and other services over the last twenty-five years, have been linked to an upward shift in London's occupational structure. Professional and managerial workers have increased as a proportion of the total resident workforce and the numbers and proportions of skilled and unskilled groups have declined. These changes in London's occupational structure are linked to changes in London's housing market which has seen a dramatic contraction of the private rented sector and the rapid growth of owner occupation and council housing. Professional and managerial demand for inner city owner occupation and rapidly rising prices have excluded the less skilled and lower paid from the owner-occupied market and concentrated them in the marginalized council sector. Competition for desirable housing is increasingly severe and this has pushed high income housing demand down market into the inner city areas traditionally occupied by the working class. There is, therefore, a trend towards the growing polarization of the population in terms of occupation, area and housing. Gentrification and residualization are occurring simultaneously in a similar way to that documented in New York by Marcuse in Chapter 4.

Housing market polarization is also occurring in Berkshire where growth of the professional and managerial labour force has placed increasing demands on the housing stock. Although there has been a substantial growth in the private housing stock, this has been largely taken up both by immigrants into the area and the upwardly socially mobile, while lower income groups have been forced to compete for the diminishing supply of lower cost housing. The result of this growing pressure has been the concentration of the less skilled workforce into what remains of the council sector and a growing polarization between highly skilled and well-paid owner occupiers and the rest. As Barlow and Savage observe, the dominance of speculative housing development in Berkshire accentuates 'divisions already opening up in the labour market, and hence helps to increase social polarization'. This has led to a contradiction between the needs of further economic growth and the inability of housing markets to provide for the less skilled. Employers report growing labour shortages and stress the need for more council housing, while the better off are determined to resist housing development and incursions into rural areas. In the medium term, the result could be either a choking off of some economic growth or its displacement into areas where the housing shortage is less acute and labour supplies are plentiful.

Hudson argues that a similar polarization process is occurring in the North-east. Whereas state policy first led to a concentration of housing development so pools of labour could be made available to employers in potential economic growth zones and in New Towns, the region's economic collapse combined with a residualist housing allocation policy led to a concentration of the surplus population into the worst estates. In consequence, these estates became, whether by accident or by design, concentrations of the economically marginalized and unemployed. At the same time, private builders were encouraged to provide owner-occupied housing to attract executives, the new middle class, and the skilled working class, particularly those households with double incomes. The intention, argues Hudson, was to transform consumption patterns and lifestyles, with the incorporation of married women into the wage labour force providing a crucial link between changes in production and consumption. Yet, while the conditions for dual wage earner owners in regular employment improved, the contrast between them and the marginalized has sharpened. The result, according to Hudson, has been increasing 'social polarization within the North-east as well as between it and the South-east'.

So, in addition to a polarization between growing and declining regions, there may be polarization within regions between those in employment in well paid jobs, and those who are trapped in poorer housing in the inner cities or council estates.

The fourth theme of the section focuses on the role of state intervention in the reorganization of labour and housing markets. Hudson argues that three phases of postwar state policy can be identified in the North-east. In the first and second phases, a restructuring of the housing market was an integral part of state policy towards the restructuring of both production and consumption relations in the region. As Hudson notes, 'The deprivation, poverty and unemployment of the interwar years was to be abolished by a state-managed sectoral and spatial reorganization of labour markets and housing provision, with public sector construction and allocation policies mediating the play of housing market forces for most of the population.' The first phase, initiated when Britain was still largely dependent on coal for energy, was designed to freeze the existing pattern of economic activities and protect the male labour force from competition, especially in coal mining. This was associated with a concentration of public housing around long-life pits and New Towns. The second phase reinforced this concentration, but encouraged owner-occupied housing as part of an attempt to create a stable labour force of dual income households. He argues that the third phase has been characterized by the abandonment of state attempts to co-ordinate changes in labour and housing markets in the region:

> – Assembling pools of labour for mobile branch plants via local authority planning and construction policies is now an irrelevance. With mass unemployment endemic throughout the North-east, companies experience little difficulty in securing labour power, wherever they locate within the region. Any pretence that the state can co-ordinate changes in the labour and housing markets has been dropped, blown away in the face of a return to generalized mass unemployment and a confirmation of the region's marginal position relative to the main currents of accumulation in the global economy (p. 235).

Goodwin's chapter also focuses on the role of state intervention in the restructuring of labour and housing markets, but the form and context is very different. Although the Docklands also witnessed the rapid decline of its traditional port-related industries, the rise of London as a global financial centre, and the proximity of Docklands to the City of London and the shortage of office space meant that it offered a prime devalorized site for redevelopment. But, while central and local state worked in tandem in the North-east, the local authorities in Docklands wanted to improve rather than replace existing housing and employment activities for local people. This was not the future envisaged by central government. In consequence, an Urban Development Corporation was set up over the heads of the local authorities to regenerate the Docklands in a way which was suitable for the new needs of financial capital.

The goal of the LDDC was not to improve *existing* homes and jobs for locals; it was to remake Docklands as a suitable site for new industry, finance capital and a luxury residential environment. Goodwin argues that the LDDC's goal of a flourishing luxury owner-occupied market has led to the replacement of an existing 'surplus population' and the area's repopulation by a new class of residents. Neither the new homes nor the new jobs were suitable for existing residents and they have largely disappeared as part of a total transformation of the area. As Goodwin notes, whereas new towns represent 'the physical embodiment of Fabian socialism and state planning, the UDCs represent the dominance of market forces'.

There are further differences between the nature of state intervention in the North-east and in the Docklands. In the North-east, Hudson argues that housing market restructuring was seen as part of the restructuring of production. In the Docklands, however, labour and housing market restructuring went hand in hand. It was necessary to shift the profile of the Docklands housing market towards owner occupation as a key part of the campaign to change both its economic and social composition. In Goodwin's words: 'Changes in the housing market and the local economy are the twin spearheads of a cumulative process of speculatively led, and financially determined development.'

There is also a difference between the form of state intervention in the North-east, where central and local government worked together in an attempt to restructure the region, and in Berkshire and the Docklands where central and local government have had very different policy objectives. The restructuring of Docklands has been imposed against the wishes of the local authorities. And in Berkshire local government has tried to slow down the pace of development which has been led by market demands for more land for housing and office space and the associated need for costly infrastructural provision. Berkshire county council faced strong opposition from central government who are keen to foster further growth in the area and have increased county housing development targets. This reflects the other side of the uneven development coin. In the North-east the state intervened to try to foster growth. In

Berkshire the local state is trying to slow it. The uneven geography of growth and decline in both labour and housing markets is thus a pervasive feature of modern Britain.

References

Bramley, G. and Paice, D. (1988), *Housing Needs in Non-Metropolitan Areas.* Association of District Councils.

Hamnett, C. (1984), The post-war restructuring of the British housing and labour markets: a critical comment on Thorns. *Environment and Planning A*, 16, 147–61.

Hamnett, C. (1988), Regional variations in house prices and house price inflation in Britain, 1969–88 *Royal Bank of Scotland Review* September, 159, 29–40.

Hamnett, C. (1989), The Owner-occupied housing market in Britain: A North-South divide? in J. Lewis and A. Townsend (eds) *The North-South Divide* (London, Paul Chapman).

Hamnett, C. (1989b), Cycling across the gap. *Housing Review* 38, 3, May–June, pp. 83–5.

Hamnett, C. and Randolph, W. (1986), 'The Role of Housing and Labour Markets in the Production of Geographical Variations in Social Stratification', in K. Hoggart and E. Kofman (eds.), *Politics, Geography and Social Stratification* (London, Croom Helm).

Hughes, G. and McCormick, B. (1981), Do Council Housing Policies Reduce Migration Between Regions? *The Economic Journal*, 91, 919–32.

Hughes, G. and McCormick, B. (1987), 'Housing Markets, Unemployment and Labour Market Flexibility in the UK', *European Economic Review*, 31, pp. 615–45.

Metcalf, D. and Richardson, R. (1980), Unemployment in Inner London, in A. Evans and D. Eversely (eds) *The Inner City: Employment and Industry*. (London, Heinemann).

Minford, P., Peel, M. and Ashton, P. (1987), *The Housing Morass: Regulation, Immobility and Unemployment*, Hobart Paperback, no 25 (London, Institute of Economic Affairs).

Steadman-Jones, G. (1971), *Outcast London* (Harmondsworth, Penguin).

Thorns, D. (1982), Industrial restructuring and changes in the labour and property markets in Britain, *Environment and Planning A,* 14, 745–63.

Vance, J. (1966). Housing the worker: the employment linkage as a force in urban structure, *Economic Geography*, 42, 4, 294–325.

6 *House price inflation and local labour market influences in Britain*

MIKE COOMBES, TONY CHAMPION &
MOIRA MUNRO

Introduction

House prices in different parts of the country have diverged over recent years, highlighted by the well documented price inflation in the South-east of England that has apparently left many other parts of the country behind. The attention that this has received is some reflection of the importance that is attached to housing generally, and the affordability of housing in particular. Divergent house price trends may also be viewed alongside the strong regional trends inherent in the profound structural shifts in employment and population over recent decades. High house price inflation, in this context, is often seen as further evidence of the markedly different economic conditions that prevail in the South-east, when compared to most other regions of the country. However, the connections between housing and labour markets are not yet well researched; previous academic work has focused almost exclusively either on housing markets or on labour markets.

Even in the limited field of house price studies, empirical research in this direction has tended to come up against the difficulty of obtaining a large national data, set on house prices with detailed geographical referencing of the individual properties. Although house price data published by building societies has become more sophisticated in recent years, this is most often disaggregated by a broad typology of housing types and by region (Fleming and Nellis, 1987). These standard statistics have formed the basis for some analysis at national and regional levels (for example, Hamnett, 1984) and have also been instrumental in bringing the issue of divergent house prices to public attention. This chapter seeks to investigate the causal processes underlying particular levels of prices, or rates of inflation, by placing the local house price 'outcome' in a context of demand and supply factors within each housing market.

The first necessity, though, is to preview the principal issues that this chapter should address. One common spatial perspective on the housing and labour markets in Britain is that both arenas are dominated by the so-called north–south divide (e.g. Brittan, 1986). While this is clearly a powerful image that is reflected in many aggregate measures (some of which will be demonstrated below) this chapter intends to concentrate on more disaggregated systems. Housing and labour market interactions can be considered at three levels of spatial disaggregation within the national economy: the individual/household level, the local housing/labour market, and finally at the regional level. To date, the majority of academic discussion has focused on the national or regional scale. In this chapter the Local Labour Market Area (LLMA) scale provides a sharper focus to identify those processes that link housing market outcomes to labour market factors.

This chapter will, first, describe some of the issues involved in developing an explanatory framework in terms of housing and labour market interactions at the local scale. The later section of the chapter will describe some of the results already available from an analysis carried out at Newcastle using data from the Nationwide Anglia Building Society. The analysis raises further questions about how to develop a more realistic understanding of the factors causing major differences in house price inflation between different parts of the country. At the local scale in particular, economic growth or decline is not the sole factor in determining housing demand, and 'excess demand' may in turn have an influence on growth prospects. It is through these local constraints that the *impacts* of the interaction between housing and labour markets begin to be revealed. The most commonly discussed impact envisages housing acting as a barrier to mobility which thus inhibits labour market clearing and, consequently, it has been argued, acts as a brake on the recovery of the national economy (Waterhouse, 1986). At the local scale, it also becomes possible to identify the broad pattern of 'winners' and 'losers', and to speculate on the impact upon local firms of immobile labour which in some areas may result in recruitment difficulties or pressure for higher wages due to housing costs. The implications of the widening divide in housing outcomes that is investigated in this chapter are considered more fully at the end.

Explanatory frameworks

The starting point here is the neo-classical mechanism by which, in both labour and housing markets, contrasts in local demand pressure flow directly out of local economic growth or decline. This would, at its simplest, explain house prices within each local area as a function of effective local demand in the area, which results from the income and demographic characteristics of the area. However, explicit consideration must also be given to interactions between localities as part of dynamic processes in housing and labour markets, as mobility by both firms and people would be expected to even out some differences between areas. There is an ample literature on the geography of industrial change in Britain, but rather less on the relationship between local industrial change and labour migration – and very little on the extent to which labour migration patterns and rates are influenced by housing factors. The pioneering work of Johnson, Salt and Wood (1974) has generated very little follow-up until recently. Hughes and McCormick (1981, 1987) have investigated the effect of the housing system on labour mobility and have argued that local authority tenants are most constrained in making long distance moves while mobility is apparently most free for private rental tenants. Research on house prices has typically focused on the national level and only rarely at a regional level. Such models focus on aggregate relationships between financial markets and the real sector (for example, Hadjimatheou, 1976) and may be embedded in full macro models (Whitehead, 1974). Research on local house price trends has typically taken a very partial view and has largely tended to restrict itself to explanations couched in terms of the physical characteristics of the individual properties and, in some cases, their neighbourhoods. Little account has been taken of local labour market conditions or, indeed, of the characteristics of the people buying into a particular area (Bourne, 1981). It is therefore appropriate to start with a justification for adopting the local scale when developing an explanatory framework.

The local scale in housing and labour markets

It is not difficult to illustrate ways in which housing markets do indeed have a strong local dimension. Individual preferences about residential location tend to focus on a particular urban system as a result of a wide variety of factors such as workplace location, closeness of family and friends, the quality of local schools and other services, and the differences in local government taxes (especially around metropolitan locations). In as much as travel to work is an important consideration for many people, it would be expected that the boundaries of the local housing market will coincide, at least to a fair degree, with local labour market areas (LLMAs) – although this remains an issue to be empirically verified. The administered non-market housing system operates within local boundaries, too, even though these local government boundaries need not, of course, be coterminous either with the boundaries of LLMAs or of private housing markets. There is also a further blurring of the local dimension to the examination of housing market outcomes, introduced here in the evidence that contiguous housing markets interact. These processes may be observable statistically as spatial auto-correlation, but whether price rises in one area may be said to 'spill-over' through time into rises in neighbouring areas remains unproven (Rosenthal, 1986).

Even accepting these reservations and imprecisions, the local scale of analysis is both defensible and illuminating. The exceptional experience of house price inflation around Aberdeen during the oil boom (Maclennan and Jones, 1986), is an extreme but indicative case of a general link between local housing and labour markets – rather than being the sole example of an idiosyncratic process. Equally, it is unlikely to be coincidental that the house price ripples beyond that city petered out around the limits of frequent commuting distance from Aberdeen. The obverse effect of stagnant or declining house prices has been observed across those labour market areas centred on towns suffering major industrial closures. Elsewhere, the linkages are less evident due to the patterns of change being less dramatic – in these areas the effect is less noticeable because it is more incremental, not because the processes are very different.

Theoretical linkages

The local scale of explanation is likely to draw initially on hypotheses developed at the regional and national scales, yet even here the theoretical work available to link housing and labour markets is scanty (Munro, 1986). There is, therefore, little guidance from previous work as to what are the most important labour market factors to include in an empirical investigation of local housing outcomes. Indeed, many such studies have been restricted to a single city (Wilkinson and Archer, 1973); but even where a broader regional or national sample of house transactions has been available, the labour market factor has been under-developed or ignored (McAvinchey and Maclennan, 1982). The first task is, therefore, to try and identify the associations between housing and labour market variables through developing an explanatory framework which includes a variety of labour market processes.

The model specification process needs to be both conceptually and empirically driven, as it is possible to postulate a very wide variety of linkages, none of which is obviously superior to all other alternatives. For example, a primary link in the schema is between levels of income from employment – after controlling for spatial differences in costs of living (Reward Regional Surveys, 1987) – and expressed demand in the housing market. Yet if this link is considered in more detail, it will be recognized that the actual influence of, say, the local skilled manual workers' level of income (or change in income) within one LLMA will depend on the distribution of that group across the housing market. Each income holder will respond as an individual, influenced by his or her life cycle stage and current tenure position, job security, mobility in both housing and labour markets and so forth (Salt, Chapter 3).

Thus, monitoring variations in house price inflation at the LLMA scale observes the outcome of all these micro-processes, together with meso-level influences such as the proportion of local households who are owner occupiers. This variation is mapped onto the structural effects from the macro-economy, such as the availability and cost of credit for house purchase, the relative subsidy in different tenures and other changes in the housing policy regime (Buckley and Ermisch, 1982). Although these factors apply nationally, they may all have differentiated impacts across contrasting local conditions. Housing market outcomes will be influenced by the interaction of all these (demand) factors with the area's available housing supply. Indeed, some sectors in the housing market, or the whole market, may be relatively unconstrained in some instances. Any subset of these variables may prove to act as indicators for the overall interactive process. Multiple regression type of analysis is the most straightforward way of estimating the level of association in this first phase of work in order to find a parsimonious set of independent variables that could be used in explicit modelling of the system. Interaction between the empirical work and the re-conceptualization of the explanatory framework offers the most stimulating approach.

The limited theoretical work completed to date suggests that the interactions operate in both directions; that is to say, labour markets both affect and are affected by housing market conditions (Thorns, 1982). Indeed, it is the possibility that a process of decline or expansion will build-up cumulatively, through feedback processes, that is one of the key policy concerns reflected in the debate around the north–south divide in labour and housing markets (Champion, Green and Owen, 1987). It is therefore important not only to look at levels of prices, but also to examine the rates of change in those prices. It is even more important to consider not only labour market conditions in cross-section, but also to analyse the changes that occur within the same time period. It should also be noted, however, that if such mutual interactions are possible then it becomes extremely difficult to identify accurately the underlying causal processes.

Specifying a preliminary model

The starting point for developing a local scale framework is the neo-classical model which may be summarized as 'predicting' housing outcomes from causal factors that focus on local trends in employment, plus measures of income and demographic characteristics such as participation rates and household structures. Due to the divisions within housing markets, and the stratification of labour markets, a single monolithic 'model' is unlikely to reflect the variety of processes operating in different sectors (public/private tenure types; high/low mobility groups, etc.). The approach in the remaining part of this chapter follows from an attempt to test the neo-classical assumptions within a statistical model. Initially, a 'black box' approach is adopted, with the explanatory framework to be developed as an 'enhancement' of the analysis. The objective is to assess how far this aggregate market-orientated conceptualization can cope with probable segmentation and stratification within the processes being observed (Hamnett and Randolph, 1984).

For the present purposes, therefore, the explanatory framework is essentially of a causal relationship between local labour market factors and local house price trends. The short-term adjustment process in the model assumes that a cross-sectional analysis can treat as exogenous, or insignificant, the influence of supply-side changes. Both social and demographic structures can also be taken to be too gradual in their adjustment to influence the observed housing outcome (Champion *et al.*, 1987). The local house price is presumed to reflect the balance of supply and demand across the whole local housing market, with tenure switch perhaps acting as a linkage between the observed private housing sector and other tenure groups. Given the recent divergence in house price trends across Britain, one of the more defensible assumptions in this simplified model may be that the current differences between LLMA mean house prices reflect the rates of recent inflation – that a 'snapshot' is an adequate proxy for the dynamic pattern leading up to it. Thus, the neo-classical framework has been 'stripped down' into a short-term model in which the differences in present LLMA house prices is a result of inflation over recent years which in turn reflects the demand (rather than supply) side of the whole local housing market, within which differences between areas are hypothesized as being the outcome primarily of relative prosperity in the local labour market. The actual process of this linkage remains a rather ill-specified amalgam of the relative trend in disposable incomes, and net-migration into the LLMA as a response to its economic growth or decline.

Exploratory empirical analysis

The discussion so far has introduced a wide range of hypotheses and also emphasized the reciprocal nature of the relationship between housing and labour market processes. The need to stress dynamic rather than static analyses adds yet another potential complexity to the research agenda. For the exploratory work beginning here, it is necessary to simplify the 'model' in the first instance, and to identify those factors for which *only* the dynamic perspective is important. For example, an area's population size is in itself an unlikely determinant of housing outcomes, but population growth or decline is obviously an important factor. In contrast, house prices need to be monitored both as a local trend through time and as values prevailing at a point in time. More importantly, the recent period has been one dominated by divergence in local house price inflation, so that high or low prices at the present will generally reflect a local inflation rate that has been relatively fast or slow, respectively.

The possibility of analysing local house prices cross-sectionally, and relating these patterns to labour market processes is crucial for an initial analysis at the LLMA scale because (even if there are few observations per month in some LLMAs) only a single year's data may be needed. The labour market processes need to fit within a model which allows for the possibility that different linkages operate within different types of local area. Empirical analysis may thus limit the generalizations that can aggregate across LLMA groups which apparently have similar linkages between labour and housing markets. For example, analysis of LLMAs may indicate that the north–south divide is an important element of such a taxonomy, but could also find cross-cutting divisions superimposed.

Building society data on house prices

Earlier research has been restricted by the problems of compiling a database which will allow detailed coverage of local house prices. Although building societies are the primary source for much research, even a complete record of building society data would not wholly cover the housing market. Transactions funded by cash or informal loan will not be captured, so that some sectors (in this case mainly the low price/quality sector) will be under-represented.

Similarly, the larger role played in some years by the commercial banks will lead to underestimates of lending during those periods of their major involvement (mainly in the higher priced sector). It is contended, however, that even an imperfect database will allow preliminary study as a basis for the detailed work needed at the local area scale. The Halifax Building Society have issued a standardized series using hedonic pricing techniques to take account of the different characteristics of houses sold in each time period (Fleming and Nellis, 1985a). The Halifax have not, however, yet succeeded in combining this 'mix adjustment' with detailed spatial disaggregation such as their new data series at the county and post town level (Halifax Building Society, 1987).

In 1986 the Nationwide Anglia Building Society (NABS) allowed the Centre for Urban and Regional Development Studies (CURDS) access to its records on mortgage lending (approaching 10 per cent of all building society loans) for the specific purpose of exploring the links between house prices and labour market conditions at the sub-regional scale. The integration of the analysis of house price trends with labour market trends requires a timeseries of house price data for areas that relate to local labour market areas 'on the ground' – and in the available timeseries of population and employment data. This data convergence was achieved for a set of areas that were specially designed to be a 'state-of-the-art' definition of LLMAs (Coombes *et al.*, 1982). The critical advance from earlier research was obtaining *grid referenced* house price data from NABS, as documented in Annex 6.1 (p. 187). Such a large dataset represents a major research opportunity for the study of geographical variations in house prices and for their analysis by reference to likely causative factors. In particular, it would seem to lend itself not only to the study of local property price patterns and trends, but also the characteristics and housing histories of purchasers and the way in which the various types of purchasers match up with particular types of property and how this relationship varies between places and alters over time. Only the first of these, the evidence on the spatial variations in property prices, will be discussed in detail below.

For the present, the investigation focuses on a cross-sectional analysis of house prices averaged for CURDS Local Labour Market Areas for a period of twelve months – related to measures of population change, employment trends and social well-being. The following section reports the results of exploratory work on a cross-sectional analysis, using the mean house price for 1985 calculated for each of the 280 CURDS Local Labour Market Areas. The measure relates to all the properties which were deemed by the NABS to have been transacted on the open market, i.e., excluding cases such as council houses bought by sitting tenants at a discount and houses in a poor state of repair where 'the valuation allowed for improvement' (see Annex 6.1, p. 187). The mean measure was found to be as robust as the median for individual LLMAs. The sample size allows this cross-sectional picture of house prices to be related to labour market dynamics, as discussed earlier. Later work on house price inflation across the LLMAs will need to consider the way in which these dynamic processes of population, employment, unemployment – and their local variations – have been changing over time.

Regional and local variation

In order to establish the LLMA scale of analysis, it is important to demonstrate positively that there are wide variations in house prices both between *and* within regions. The ten highest and lowest priced areas according to this calculation are listed in Table 6.1 and show an impressively wide margin between the highest and lowest priced LLMAs. Figure 6.1 shows the overall picture and indicates that the geographical patterns are generally clear cut. The peak zone of mean house prices occurs immediately to the west of the London LLMA and values fall fairly regularly in all directions from there. Particularly well marked is the £40,000 class interval which picks out the major concentration of high house price areas in the South-east, contrasting with the mere handful of similar places scattered around the rest of Britain. Table 6.2 indicates the relative position of regions in Britain on house prices – a pattern that is now quite well known from the regular bulletins issued by certain building societies as well as from the Department of Environment's 5 per cent sample of building society mortgage completions (Fleming and Nellis, 1985b). This shows that the South-east is in a class of its own, the South-west and East Anglia come next and are fairly similar to each other, and the differences between the remaining regions level out within a relatively narrow range.

Much less well known than this broad regional pattern, however, is the extent of the variation in mean house price between LLMAs *within* each region. As shown in Table 6.2, this amounts to a difference of £33,000 in the South-east, where the mean house price for Deal is only half that of Guildford. Even in the lower priced regions the range does not fall below £18,000 – not much less than the difference in mean price between the highest and lowest Standard Regions. This previously little known LLMA scale in house price variations does not, of course, contradict the well established neighbourhood scale of house price differentials – the 'right' or 'wrong side of the tracks' for residential location within a town

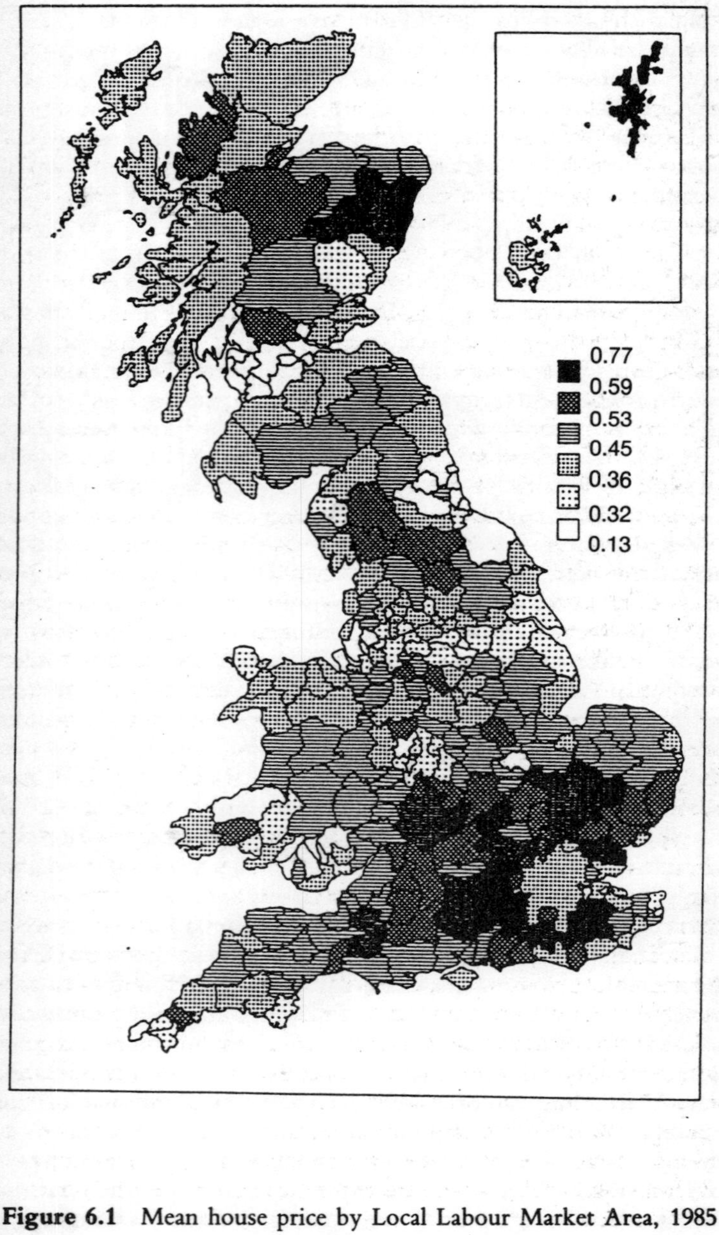

Figure 6.1 Mean house price by Local Labour Market Area, 1985

Table 6.1 Mean house prices 1985: extreme LLMAs

Highest	£	Lowest	£
Guildford	66,488	Consett	18,689
High Wycombe	65,159	West Bromwich	20,786
Maidenhead	63,039	Hartlepool	20,952
Woking & Weybridge	61,984	Mansfield	22,295
Bracknell	61,767	Goole	22,547
Watford	60,617	Wrexham	23,171
Welwyn	58,708	Mexborough	23,207
St Albans	58,196	Neath	23,470
Tunbridge Wells	57,757	Barnsley	23,527
Hertford & Ware	55,625	Holyhead	23,705

Source: calculated from data provided by the Nationwide Anglia Building Society. The data relate to market-price transactions for which the NBS approved mortgage advances in 1985.

or city. Given detailed location coding on the house price dataset this dimension needs to be disentangled from the LLMA analysis and some possible strategies are included in the discussion of future work outlined later, along with the need to consider the fourth level of differentiation, between individual houses in a neighbourhood, which reflects house size and characteristics (and so is not really a spatial factor as such).

Local house prices and labour market conditions

It is now appropriate to move towards investigating the labour market influences that, through differential house price inflation, are hypothesized as part of the explanations for these spatial differences in housing market outcomes. Both the overall regional pattern of house price differentials and the types of places with highest and lowest mean house prices within regions, outlined above, may be seen as confirming the importance of labour market forces in determining the geography of house prices. A simple atheoretical way of exploring this relationship initially is by comparing mean house prices with the Index of Local Economic Performance developed by Champion and Green (1985). The economic fortunes of Britain's 280 LLMAs are summarized by this analysis, now more widely known as the Booming Towns Index. The aggregation of factors that are 'bundled' together in the Booming Towns Index comprises three indicators of population and employment change, plus the unemployment rate and, as a surrogate for affluence, the proportion of households with high car ownership. All these variables were among the range of factors in the preceding discussion of the analytical framework for this study, they were selected for the Index because they were readily measurable at the LLMA scale.

The Index score for each LLMA is mapped in Figure 6.2, the patterns revealing a strong degree of similarity to the geography of house prices both in general and in various particulars. Thus, Consett is not only the lowest priced housing area according to the NBS's 1985 data but is also bottom of Britain's 280 places on this Index, with Hartlepool and Mexborough joining it from Table 6.1 as being among the lowest ten LLMAs by both house prices and Booming Towns Index. At the other extreme, seven of the ten highest priced LLMAs in Table 6.1 also feature in the 'Top Twenty' Booming Towns, while Aberdeen stands out distinctly on both Figures 6.1 and 6.2 as having relatively high house prices and a strong economic performance. The zone of very high average house prices immediately west of London coincides with the most dynamic LLMAs, while the poorer economic performance of LLMAs east of London along the Thames estuary is reflected in the considerably lower house prices there despite the opportunities for commuting to London. The broad north–south divide is very evident in both cases, with a particular correspondence in South-east England between LLMAs with an average house price

Table 6.2 House prices 1985 by standard region: means and extreme LLMAs

Region	Mean price for region (weighted) £	Highest LLMA	£	Lowest LLMA	£
South-east	49,592	Guildford	66,488	Deal	33,740
South-west	38,848	Salisbury	45,066	Redruth & Camborne	27,457
East Anglia	38,369	Cambridge	47,870	Lowestoft	27,580
West Midlands	33,238	Stratford	47,140	West Bromwich	20,786
Scotland	33,228	Stirling	43,880	Kirkcaldy	25,216
North-west	31,232	Macclesfield	44,073	Accrington	24,414
East Midlands	32,128	Newark	40,294	Mansfield	22,295
Northern	30,653	Hexham	48,640	Consett	18,689
Wales	30,245	Monmouth	41,424	Wrexham	23,171
Yorkshire & Humberside	29,783	Harrogate	42,163	Goole	22,547

Source: Calculated from data provided by the Nationwide Anglia Building Society.

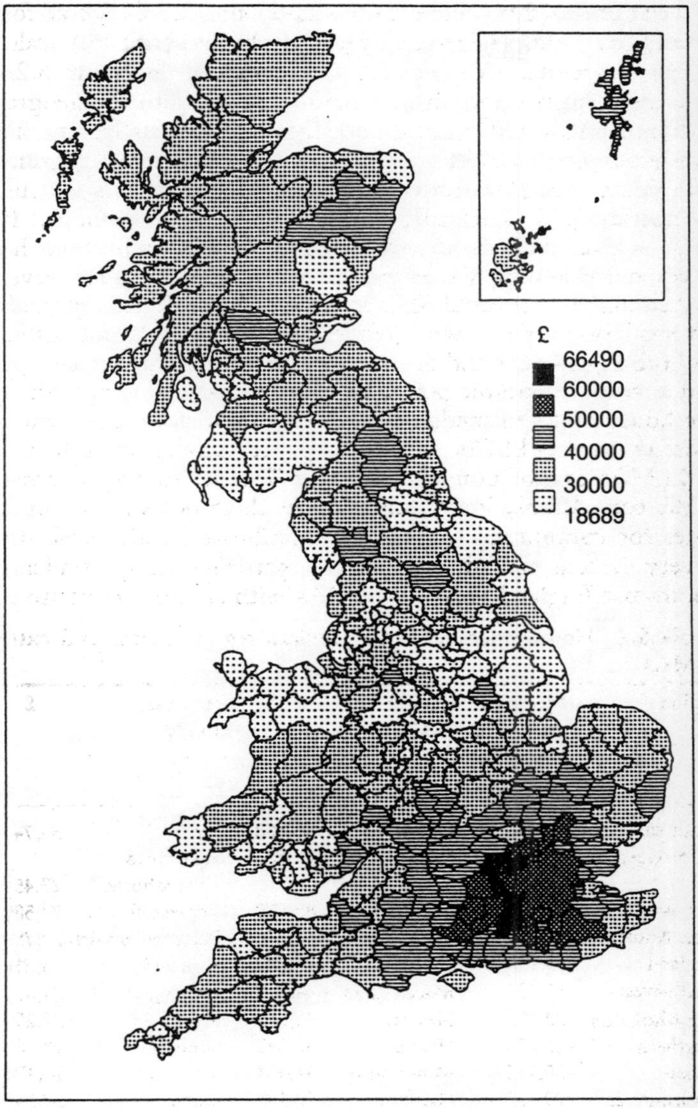

Figure 6.2 Index of Local Economic Performance by LLMA

of over £40,000 and those with a Booming Towns Index of 0.53 and over.

The relationship is confirmed as a strong positive association between house prices and economic performance by a correlation coefficient (r) of 0.74, giving a 54.8 per cent level of explanation of variance in regression analysis. Examination of the residuals from the best-fit linear relationship (Figure 6.3 and Table 6.3) indicates a strong London effect, with both London itself and a number of the surrounding LLMAs being characterized by much higher house prices than predicted from the overall relationship. A secondary concentration of high positive residuals is found in west-central Scotland around Clydeside, where house prices are not as low as the very poor economic performance there would suggest. By contrast, the places where house prices are lower than predicted on the basis of their Booming Towns Index are much more varied in location and character – perhaps 'good territory for the bargain hunter' in the jargon of the property industry commentators or, alternatively, places suffering from disadvantages which are not captured by the Index in this form.

Further developments

The evidence of this very preliminary analysis indicates the strong general association between housing and labour market conditions, but there is clearly need for more work to expose the underlying mechanisms more clearly. The remaining parts of this chapter discuss three broad fronts on which the exploratory work above can be advanced. First, the initial regression analysis can be advanced by 'unpacking' the Booming Towns Index to investigate the significance of individual component variables reflecting the earlier discussion of housing and labour market interactions. Second, the mean 1985 NABS house price needs to be replaced by more sensitive and dynamic measures of housing market outcomes. Finally, the empirical work must stimulate further development of explanatory frameworks which in turn provide better model specifications of labour and housing market interactions at the local level.

Enhancing the local labour market indicators

The Booming Towns Index incorporates five LLMA indicator variables, each of which need to be considered separately as potential influences upon housing market outcomes, as a step towards a less 'black box' analysis. The 'prediction' of house price variation can

Table 6.3 Standardized residuals from the best-fit linear relationship between mean house price and Index of Local Economic Performance: the extreme LLMAs

Largest positive		Largest negative	
Guildford	3.44	Mansfield	−2.07
Watford	3.15	Carmarthen	−1.93
London	2.91	Peterhead	−1.89
Welwyn	2.88	Goole	−1.79
High Wycombe	2.76	Stafford	−1.76
Harlow	2.70	Wrexham	−1.71
Woking & Weybridge	2.53	Redditch	−1.70
Gravesend	2.36	Bangor	−1.70
Maidenhead	2.36	Heanor	−1.66
Tunbridge Wells	2.31	Coalville	−1.65

Source: Author's booming towns index.

therefore be developed into multiple regression by taking these individual variables along with other labour market indicators of those factors discussed in the theoretical review. Accordingly, Champion and Brunsdon (1988) detail twenty-two independent variables in a regression analysis that deals with the extreme values in the house price dataset by taking logarithms of the LLMA means. A number of analytical variants were explored but the main findings were robust. First, the single variable that was consistently most strongly correlated (positively) with the house price measure was the proportion of the LLMA's 1981 workforce who were professionals or managerial. The next five most influential factors included three from the Booming Towns Index – unemployment rate (negatively correlated), employment change 1978–81 and the car-ownership affluence surrogate (positive). The other two consistently influential factors were the share of the LLMA's households who were owner occupiers and the proportion of workers employed in the finance sector (which is nationally the most dynamic sector for job growth).

Together these six variables in a stepwise regression analysis statistically account for over 75 per cent of the variation in the LLMA house price measure. More importantly, they empirically confirm the influence of the range of labour market factors set out in the earlier review. The hypothesized importance of local employment growth or decline is confirmed, though here it is focused on the recession period. More recent trends are represented here by the focus on a particular growth sector, while local unemployment levels are independently influential and the importance

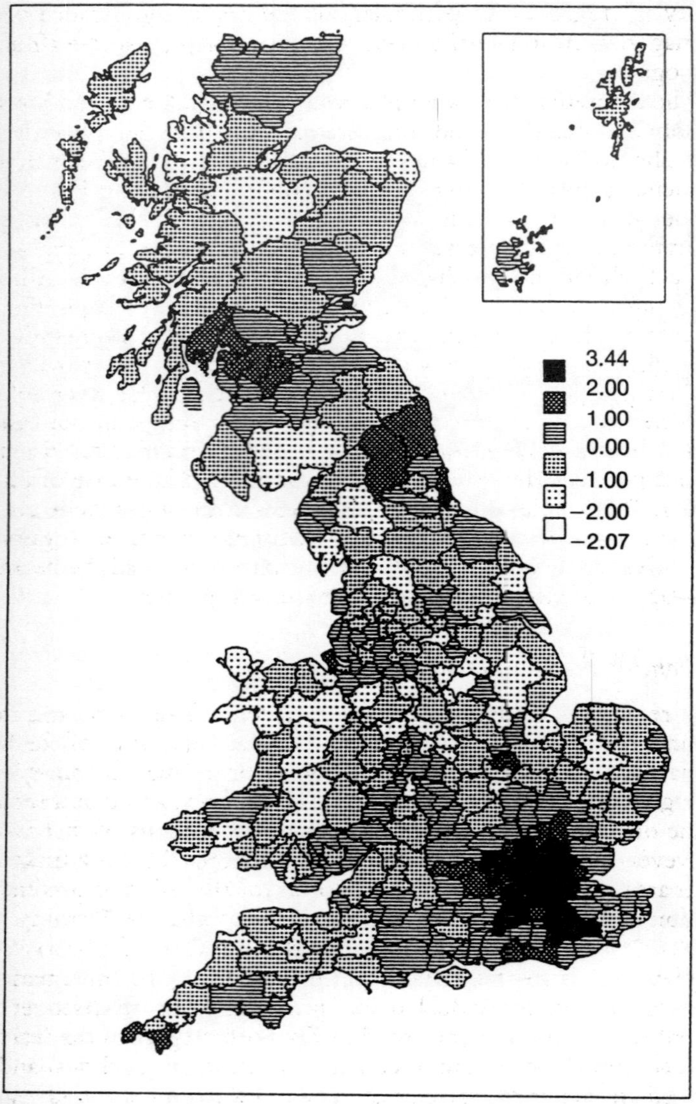

Figure 6.3 Residuals from best-fit linear relationship between mean house price and Index of Local Economic Performance

of the local tenurial structure is reflected in the owner-occupation variable. More speculatively, the importance of segmentation in the labour market is one possible interpretation (given the selectivity of migration in particular) of the major significance of the managerial/professional workforce and, perhaps, also the affluence surrogate.

The theoretical review emphasized that the link between housing supply and demand is not uni-directional, so that supply-side factors should be part of a model alongside labour market variables influencing local housing demand. Supply of housing is expected to be relatively inelastic in the short-run, but when looking at evolving systems over several years the possibility of supply-side impacts should not be dismissed. Consequently, analysis of house price *inflation* will need to measure changes in the public and private stocks of housing within areas, including consideration of demolitions, building activity and local authority sales (particularly under the right-to-buy, introduced in 1980). Such changes might be expected to operate *directly* on the price of housing in the case of private sector stock changes, and *indirectly*, through altered tenure choice (perhaps due to changing accessibility) in the case of public sector. While this will be an improvement on work done so far, it presents only a partial picture of supply response – ignoring the privately rented sector entirely. These issues might be more appropriately handled within a case study approach.

Enhancing the housing outcomes measures

The regression analyses above have all been based on the simple mean LLMA house price measure. Even as a measure of the local private housing market, this is clearly a 'primitive' indicator, even though comparative analyses provided confirmation that a median value offers no notable enhancement to the analysis. Other work, however, has stressed the influence of different 'housing mixes' in each area, so that some standardization for this effect is preferable. Unfortunately, hedonic procedures such as that of Fleming and Nellis (1985a) are not feasible on the number of observations available for many LLMAs. The need here may be for a national investigation of individual house prices in order to discover the extent to which these vary predictably with respect to the features of each house (type, date of construction, rooms, facilities) and its immediate geographical environment. This could be undertaken for a full sample of houses for a given period (say twelve months), but could also be done for sub-groups of houses (as in the Halifax Building Society (1987) Post Code Area analysis of three-bedroom semi-detached houses built since 1945 but not new). Similarly, in examining rates of change in house prices, differing mixes of houses sold or improvements in the average quality of housing sold (either through the impact of new build or from upgrading of the existing stock) would affect rates of house price increase.

Measures of the geographical environment involve variables relating to the socio-economic composition of the local residential neighbourhood. The neighbourhood is a focus which seeks to capture the aspect of purchasers' perceptions of different areas that is reflected in estate agents' jargon as 'premium' or 'bargain' addresses within particular cities. Although it is easily understood as a generality, it is a scale of variation that is difficult to operationalize for objective analysis. Clearly, the crudest level may be the oft cited distinction between East End and West End of many cities, or between suburbia and the blight (or gentrification) of inner cities. Once the locally distinctive pattern of public sector housing in each town has been set aside, even these simple sub-divisions result in a quite complex pattern of locales which (as shown by Munro and Maclennan, 1987, in Glasgow) do appear to experience different house price trends. The alternative is to interpret neighbourhoods not as broad tracts within any local housing market, but as streets and addresses (across the whole country) which have similar socio-demographic profiles (see Openshaw, 1987). In this way, house price analysis could monitor together all 'high status' addresses across a city – rather than analysing the broader high status locale that will include some lower priced enclaves within *any* sort of definable boundary for Surbiton or Solihull, Bramhall or Bearsden.

Once a better LLMA indicator has been devised, the need is to move on to dynamic analysis of house price inflation. Time series analysis will need to test for the existence of an autocorrelative process, given the upward trend of prices through the period in many areas; this would be a further restriction given that the number of independent observations is already reduced. In any case, to disaggregate the time series further (say into a quarterly series) would probably result in small numbers of observations per area and hence unreliable estimates. It would also introduce the extra complicating element of seasonal variations: there is evidence in Glasgow, for instance, of a recurring pattern of price rises through the year – reflecting the levels of turnover which peak in the spring and summer and fall back through the other two seasons. Local inflation can be measured by estimating difference equations – that is to consider the changes between one year and the next in relation to changes in values in the key factors. This is a strong technique for abstracting from the absolute levels of different factors and concentrating on the analysis of change.

Enhancing the explanatory framework

The exploratory nature of this chapter has justified a highly reductionist approach to the influences on local housing market outcomes. Although owner occupation is now the dominant tenure nationally, in many LLMAs it is inadequate to take the house price time series as an indicator of supply and demand pressures generally across all tenure types. The stratification of housing markets needs to be handled explicitly, with the dynamic analysis recognizing local variations in 'tenure switch' behaviour in the recent period of radical shifts in policy regimes that have hastened the decline of the public rented sector and set the scene for a revitalization of private renting.

On the labour market side of the analysis, particular attention may be needed to clarify the impact of migration, which is central to the neo-classical model of local housing demand differentials. This in turn emphasizes the issue of labour market segmentation, for longer distance migrants are predominantly in the higher paid occupation groups (Forrest and Murie, 1987). As a result, the influence of labour market pressures may not extend strongly across the whole housing market, with demand for houses in the 'first-time buyer' bracket perhaps more influenced by the local demographic structure and household formation rates. Some studies have emphasized this factor at the national level (for example, Kleinman, 1988) but once again its transferability to the local scale requires careful elaboration. Finally, to the extent that demand for housing is influenced by the value of housing as an investment, then past inflation differentials may be self-perpetuating, given all other factors being constant.

Conclusions and implications

This chapter has focused on the potential for shifting analysis of labour market influences upon housing outcomes down to the local level. The empirical work has demonstrated both the feasibility of a local perspective and the insights obtainable, while also indicating the clear prima-facie relationship with local labour market conditions. Further work to draw out this range of influences, between the neighbourhood and the regional/national levels, seems likely to elucidate many processes that so far have been consigned to one 'black box' or another.

Many important theoretical and empirical issues remain to be addressed in the complex field of the interaction of housing and labour markets. At the same time, there are many important issues of equity between households and areas that arise out of the large differentials in housing market outcomes at the LLMA scale. Those households wishing to move into high priced areas for work will find it hard to afford a house (and possibly hard to sell their previous house) although owner occupation is often the only feasible tenure option for long distance migrants (Green *et al.*, 1986). Council house allocation procedures typically give local needs a high priority, making it difficult for longer distance migrants to find a place in this hard-pressed sector. Even home owners who wish to move out of a high priced area will be deterred if they expect to have to return at some future date, as they might be reluctant to lose their foothold in the higher priced area.

These factors will have real impacts on the ability of firms in particular LLMAs to recruit people into their workforce, and the consequent strategies (of increasing help with housing and relocation costs directly and increasing 'cost of living allowances' to workers) are likely to increase firms' costs and also to fuel still further accelerations in house prices. Thus recruitment difficulties pertain on either side of a local house price divide, although in different sectors and for different types of worker (Parsons, 1987). These factors are likely also to affect firms' location decisions, so that even if they anticipate finding cheaper accommodation (and perhaps cheaper labour) outside the booming areas, they would be discouraged from moving by the potential difficulties of recruiting particular types of staff. Although there are clearly many other important influences on firms' location, factors such as this may explain in some part the apparent failure of the labour market to equilibriate inter-regionally (McCormick, 1987). Some of the inter-LLMA rigidities probably underlie the substantial growth recently in longer distance commuting.

There are other consequences of spatial house price differentials which flow from the housing finance system, but are less commonly discussed. At the household level there are important issues of equity which arise both in relation to differences in levels and also in rates of appreciation of house prices. Owner occupiers currently receive tax relief on their mortgage (up to a maximum limit of loan at their highest income tax rate) and this constitutes a greater benefit to those with larger loans and who pay higher rates of tax. Further, house price inflation constitutes an untaxed capital accumulation which clearly also constitutes a greater benefit to those whose house demonstrates a greater rate of house price appreciation. Thus, even if the tax benefits from mortgage interest relief (or as some would argue the subsidy that arises from the absence of a tax on imputed income) are entirely capitalized into higher house prices, the benefits from capital accumulation would simply be enhanced. Disproportionate benefits from the tax system, then, flow to people in LLMAs that have high prices and high rates of inflation and to the extent that this co-exists with favourable conditions in the labour market this benefit also will go to individuals and areas that are relatively prosperous.

Thus, in summary, many of those who live in LLMAs with low house prices and depressed economic conditions are in fact doubly disadvantaged. As Thorns (1982) argues, they have poor labour market prospects and their house, which is their major asset, is accumulating value at a slower rate than houses in other locations. This may be offset, for some households, by their ability to afford better quality housing more suited to their needs. However, this is in sharp contrast to high priced areas, where there are advantages to be reaped from the tax system and from house price inflation, but where problems of accessibility and suitability of housing are likely to be greater. For these and other reasons, there is considerable value in trying to explore which factors dictate particular housing market outcomes. This chapter has aimed to make a start in the identification of locally based processes and in describing the relationship between the local, regional and national trends in those factors.

Annex 1

The size and nature of the NABS dataset can be indicated by reference to the 1985 data tape. This contains 126,277 cases of mortgage approval or completion in relation to house purchases in the UK. Of these 94,140 (75 per cent) involved approvals made by NABS in the calendar year 1985, with the remainder mainly made in 1984 but not completed until 1985. The 'approval' stage, which refers to the initial decision to grant a mortgage, provides the most up-to-date information on trends, while also avoiding the inconsistency in 'completion' data due to the processing time that varies considerably between transactions. Of the 94,140 cases, 71,077 (76 per cent) were deemed to be at market price, with the most common reasons for transactions at below market price being 'valuation allows for improvements' (5 per cent of total sample) and 'council tenant' (5 per cent, presumably most under the Right to Buy Scheme).

The most useful and complete items of information in the dataset fall into three groups; the property, the purchaser and the sale. The first relates to the locational and physical attributes of the property itself. The geographical location is indicated for all cases by a Local Authority District code and a one-kilometre Grid Square reference, while the full postcode is provided in about two-thirds of the cases. The very detailed geographical referencing provided by the postcode of the property was not introduced until the 1985 tape and even in 1986 this item has not been completed in over a quarter of the cases. The other characteristics are mainly concerned with the type and size of the property and year of construction, garaging facilities and central heating system add to the overall picture.

The remaining information relates mainly to the purchasers and gives details of their demographic and socio-economic circumstances. Details on age, sex, marital status and number and age of dependent children permits relatively sophisticated analysis by family type and life cycle stage, while a broad classification into occupational groupings provides useful extra insights when read alongside statistics on income. Studies of migration are, however, hampered particularly by the absence of a detailed indication of the location of the previous address. Finally, the set of variables relating to the sale itself provides the normal basic information on sale price and its split between cash deposit and mortgage advance.

Newcastle University currently holds the NABS records for the calendar years 1980 to 1986 inclusive. Generally, the numbers of cases has increased over the years, reflecting the growth in mortgage transactions during this period. Thus, while in 1980 the NABS approved 71,411 approvals in Great Britain, the figure for 1986 was 129,387.

The authors acknowledge the inputs of Dave Owen and Anne Green to the processing and analysis of the NABS data.

References

Bourne, L. S. (1981), *The Geography of Housing* (London, Edward Arnold).

Brittan, S. (1986), 'The great housing barrier', *Financial Times*, 27 November.

Buckley, R. and Ermisch, J. (1982), 'Government policy and house prices in the United Kingdom: an econometric analysis', *Oxford Bulletin of Economics and Statistics* 44, pp. 273–304.

Champion, A. G. and Brunsdon, C. (1988), 'The variation of house prices between local labour market areas: a preliminary analysis of Building Society data', *Discussion Paper 90*, CURDS, University of Newcastle upon Tyne.

Champion, A. G. and Green, A. E. (1985), 'In search of Britain's booming towns', *Discussion Paper 72* CURDS, University of Newcastle upon Tyne.

Champion, A. G., Green, A. E. and Owen, D. W. (1987), 'Housing, labour mobility and unemployment', *The Planner* 73, pp. 11–17.

Champion, A. G., Green, A. E., Owen, D. W., Ellin, D. J. and Coombes, M. G. (1987), *Changing Places: Britain's Demographic, Economic and Social Complexion* (London, Edward Arnold).

Coombes, M. G., Dixon, J. S., Goddard, J. B., Openshaw, S. and Taylor, P. J. (1982), 'Functional regions for the population census of Great Britain' in D. T. Herbert and R. J. Johnson (eds), *Geography and the Urban Environment, Progress in Research and Applications* 5 (Chichester, John Wiley).

Fleming, M. C. and Nellis, J. G. (1985a), 'The application of hedonic indexing methods: a study of house prices in the United Kingdom', *Statistical Journal of the United Nations*, ECE 3, pp. 249–70.

Fleming, M. C. and Nellis, J. G. (1985b), 'House-price statistics for the United Kingdom: a survey and critical review of recent developments', *Environment and Planning A* 17, pp. 297–318.

Fleming, M. C. and Nellis, J. G. (1987), *Spon's House Price Data Book* (London, Spon).

Forrest, R. and Murie, A. (1987), 'The affluent home owner: labour market position and the shaping of housing histories', in N. Thrift and P. Williams (eds), *Class and Space* (London, Routledge and Kegan Paul).

Green, A. E., Owen, D. W., Champion, A. G., Goddard, J. B. and Coombes, M. G. (1986), 'What contribution can labour migration make to reducing unemployment?', in P. E. Hart (ed.), *Unemployment and Labour Market Policies*, (Aldershot, Gower).

Hadjimatheou, G. (1976), *Housing and Mortgage Markets* (London, Saxon House).

Halifax Building Society (1987), 'Halifax House Price Index', *Regional Bulletin No 15*, 3rd quarter (and County Supplement).

Hamnett, C. (1984), 'The post-war structuring of the British housing and labour markets: a critical comment on Thorns', *Environment and Planning A* 16, pp. 147–61.

Hamnett, C. and Randolph, W. (1986), 'The role of labour and housing markets in the production of geographical variations in social stratification' in K. Hoggart and E. Kofman, *Politics, Geography and Social Stratification* (London, Croom Helm), pp. 213–46.

Hughes, G. A. and McCormick, B. (1981), 'Do council housing policies reduce migration between regions?', *Economic Journal* 91, pp. 918–37.

Hughes, G. A. and McCormick, B. (1987), 'Housing markets, unemployment and labour market flexibility in the UK', *European Economic Review*, pp. 615–45.

Johnson, J. H., Salt, J. and Wood, P. A. (1974), *Housing and Labour Migration in England and Wales* (Farnborough, Saxon House).

Kleinman, M. (1988), 'A medium term forecast of housing output in Great Britain', paper presented at ESRC Urban and Regional Economics Seminar Group, Glasgow, January.

McAvinchey, I. D. and Maclennan, D. (1982), 'A regional comparison of house price inflation rates in Britain, 1967–76', *Urban Studies* 19, pp. 43–57.

Maclennan, D. and Jones, C. (1986), 'The impact of north sea oil development on the housing market', *Land Development Studies* 3, pp. 113–26.

McCormick, B. (1987), 'Housing and Labour Mobility', *Employment Institute Economic Report* 3.

Munro, M. (1986), 'Housing and labour market interactions: a review', *Discussion Paper 12*, Centre for Housing Research, University of Glasgow.

Munro, M. and Maclennan, D. (1987), 'Intra-urban changes in housing prices: Glasgow 1972–83', *Housing Studies* 2, pp. 65–81.

Openshaw, S. (1987), 'Analysing and exploiting client data by creating a geographic marketing information and modelling system', *Northern RRL Research Report 2* CURDS, University of Newcastle upon Tyne.

Parsons, D. (1987), 'Recruitment difficulties and the housing market', *The Planner* 73, pp. 30–4.

Reward Regional Surveys (1987), *Cost of Living Regional Comparisons September 1987*, Reward Regional Surveys, Stone, Staffs.

Rosenthal, L. (1986), 'Regional house price interactions in the UK, 1975–81: a cross-spectral analysis', *Applied Economics* 18, pp. 1011–23.

Salt, J. (1985), 'Housing and labour migration'. Paper presented at Conference on Housing and Labour Market Change December 12–13, London.

Thorns, D. C. (1982), 'Industrial restructuring and change in the labour and property markets in Britain', *Environment and Planning A* 14, pp. 745–63.

Waterhouse, R. (1986), 'Moving south – so multiply by three', the *Sunday Times*, 2 November.

Whitehead, C. M. E. (1974), *The UK Housing Market: An Econometric Model* (Farnborough, Saxon House).

Wilkinson, R. K. and Archer, C. A. (1973), 'Measuring the determinants of relative house prices', *Environment and Planning A* 5, pp. 357–67.

7 *Labour markets, housing markets and social restructuring in a global city: the case of London*

CHRIS HAMNETT

Introduction

In the introduction to this section we argued that one of the key characteristics of housing and labour markets is that they are spatially constituted. This inherent spatiality has a number of consequences, one of the most important of which is that individual labour markets can be differentiated according to their position in the spatial organization of production and the spatial division of labour (Massey, 1984). But housing markets are also spatially constituted, as a result of both the spatially differentiated nature of the fixed housing stock and the spatial nature of demand given, among other things, by the structure and characteristics of the paid labour force within an area and the spatial limits on commuting between home and work. I stress this because some housing markets, in Florida, California, and elsewhere, primarily cater to the needs of the retired and cannot be easily related to the spatial division of labour as such. Also, some local housing markets such as London, New York, Los Angeles and Paris are strongly influenced by demand from the international rich. We cannot, therefore, simply classify housing markets in terms of the role or position of the local labour market(s) in the spatial division of labour. It is more appropriate to classify housing markets in terms of their role and position in the *spatial division of housing consumption*. This is related to the spatial division of labour, often very closely related, particularly in the case of isolated mining towns and the like, but it is not identical to it and nor can it be simply read off from it or reduced to it. Although for many people consumption of housing takes place within commuting distance of their place of work, this is not a necessary condition of housing consumption.

Given the variation in the position of different areas within the spatial division of labour and consumption, this chapter focuses on the relationships between labour and housing markets in London, a global city which plays a dominant role in both the global and national division of labour and in the global division of consumption. The focus will be on the relations between the major changes which have taken place within the last thirty years in the structure of employment, occupation and incomes, and the equally dramatic changes in the social and spatial structure of the housing market. I will argue that although these changes in the labour and housing markets have not been causally linked in the sense that changes in the structure of employment and income produced or caused changes in the structure of housing tenure, they have been very closely related, particularly through the price mechanism, and have produced a city which is increasingly polarized: economically, socially and spatially.

Variants of the polarization thesis have been argued with increasing frequency in recent years regarding a number of different global cities: for New York by Sassen-Koob (1984), Marcuse (1986) and Smith and Williams (1986); for Los Angeles by Soja (1986); and for London by Hamnett and Randolph (1986, 1988). The thesis comes in several parts. The basic economic thesis is that the changes in the international division of labour have led to a division of functions, and an urban hierarchy within which there is a growing concentration of economic and financial command and control functions in a small number of global cities which effectively dominate the global economy. These cities include London, New York and Tokyo and a number of other major cities such as Paris, Los Angeles, San Francisco, Vancouver and Sydney, some of which are more important at the national or regional scale than the global level.

This concentration is, it is argued, linked to a shift from manufacturing into services, particularly financial and corporate producer services, and a shift in occupational structures which involves the growth of professional, managerial and technical categories and, arguably, of less skilled service workers at the expense of the skilled manual working class. These occupational shifts are also linked to changes in income distribution which reveal a widening gap between the well paid 'service class', and the poorly paid 'servicing class'. The changes in occupational and income structure are related to differences in housing market power, and to a growing social and spatial polarization within the housing market with the growth of a luxury housing market for the wealthy at the top and a residual housing market at the bottom for the economically and socially marginalized. Gentrification at the top is linked to residualization at the bottom. Smith and Williams (1986) have labelled this process 'social Manhattanization'. This was captured in a striking form by Tom Wolfe in *Bonfire of the Vanities*, his novel of the divisions of class and race in contemporary New York where the new Masters of the Universe – the Wall Street traders – live in gilded affluence: a different world to the poor black underclass, and even to the marginal middle classes, maintaining a precarious toe-hold in Manhattan.

This is a general thesis, and there are obviously local differences depending on the structure of the employment and housing markets and the nature of the changes which have taken place. Thus, in Tokyo, rampant land speculation has resulted in house prices which now bear little or no relation to people's earning capacity. The house price/income ratio has risen from about eight in the mid 1980s to eighteen in the late 1980s and the division of the population into a housing rich and a housing poor group (*The Economist*, 9 September 1989). In New York, the out migration of whites from areas such as the South Bronx and their replacement by poor blacks and associated landlord disinvestment has led to massive housing abandonment in some inner city areas (Marcuse, 1986). In London, polarization in employment, occupation and incomes has taken place within the context of a rapid decline of the private rented sector and its replacement by a large public rented sector and a small but growing owner-occupied sector. Professional and managerial workers have thus become increasingly concentrated in the highly expensive inner city owner-occupied market while the less skilled, lower paid and some minority groups have been increasingly concentrated in the council sector. In arguing that the changes are closely related, I am *not* arguing that the changes in the structure of the housing market have reflected or been caused by changes in the employment or occupational structure, but that the two sets of changes have been directly related, particularly through shifts in the demand for and supply of housing and the operation of the price mechanism which has led to a concentration of the more highly skilled and highly paid in the best housing.

The general thesis is, therefore, that changes in the spatial division of labour are linked to changes in the urban hierarchy, and that in the global financial cities at the top of the hierarchy, there is a sharp polarization in occupational and income composition which is linked to a polarization in the housing market and social segregation, but that the precise form of these changes are locally contingent. The structure of the chapter follows this general argument. The first section sets out the general argument regarding the role of global cities within the changing international and national division of labour. The second section examines the changing employment structure of London, the third the changing occupational and income structure in global cities. The fourth section examines the links between the structure of occupations and incomes and the housing market, and the fifth the geographical dimensions of the housing and social polarization of London.

Global cities and the changing national and international spatial division of labour

The last few years have seen a recognition of the importance of what Massey (1984) has termed the 'spatial divisions of labour' to an understanding of the geography of economic activity. The essence of this concept is that the changing technical division of labour between different countries, different sectors of employment, and within firms has a necessary and inescapable spatial dimension in which different types of production process, and different parts of the labour process take place in different places. The spatial differentiation of economic activity is not a new idea, of course. What is new is a recognition of the radical changes in the division of labour which have taken place over the postwar period. The rise of manufacturing industry was frequently linked with regional specialization of production by industrial sector. But, as Doreen Massey (1978, 1984) has argued, this has tended to be replaced or overlaid in modern capitalist economies by a division in which different parts of the production process take place in different areas depending on their relative advantages. Thus there is a sharp difference between the location of corporate headquarters, research and development, and different types of production. Headquarter offices are generally located in the large financial centres or capital cities, whereas production is often located in cheap labour areas in the economic periphery. These give rise to specific national divisions of labour. This argument has also been extended to include the service sector.

With the rise of multinational companies, the spatial division of labour has taken on an international dimension, with corporate headquarters located in one of the major financial centres, other regional offices in major cities scattered across the globe, and production decentralized either to cheap first world locations or to the Third World. The older international division of labour, whereby Western countries exported manufactured goods in return for primary products, has been replaced by what is often termed a New International Division of Labour in which decentralized manufacturing production is controlled from the metropolitan core.

These changes in the national and international spatial division of labour have been linked with the shift, in Western capitalist countries in particular, from manufacturing to services. It is not proposed here to attempt to precisely define services (see Allen, 1987, 1988). Suffice to say, that as the processes of design, production, distribution, advertising and marketing have become more complex, so there has been a switch in employment from goods production to ancillary areas. As Hall (1966, p. 9) puts it: 'for every producer of goods, more and more people are needed at office desks to achieve good design, to finance and plan production, to promote efficient national and international distribution.' And, with the growth of professional and managerial and technical occupations, there has also been a sharp rise in financial, legal and other services for companies large and small. At the same time there has also been a growth in personal services in areas such as hotels, catering, leisure and retailing.

But, the growth of what is termed 'producer service' employment has not proceeded evenly across the globe. Instead, it is particularly concentrated in the largest financial and commercial centres such as New York, London and Tokyo. The new corporate demand for banking, law, accountancy, advertising and some other services has been associated with the rise of a series of 'global cities' which act as corporate control and command centres for the new international division of labour. In these cities the key financial and business services firms are taking a major role in advising on corporate policy and decision and image-making (Cohen, 1981).

The corollary of the centralization process is a restructuring of the urban hierarchy. The changes in the spatial division of labour and production have been reflected by changes in the structure of the urban hierarchy. Smith and Williams (1986 p. 208) argue that: 'the continuing development of a new international division of labour will be the major determinant, at the regional scale, of urban restructuring'. With the rapid re-organization of productive functions both nationally and internationally, a new hierarchy of urban functions is emerging:

> If the late feudal city was defined by the concentration of commercial capital (in 'market towns'), and the early capitalist city by the concentration of productive capital in industry, the late capitalist city is defined . . . by . . . the concentration of money capital and the gamut of financial, administrative and profession services that lubricate the money flow. It is this function that defines the late capitalist cities at the top of the new urban hierarchy (ibid, p. 211).

At the same time as the position of the handful of new global cities has been reinforced, and become part of the international financial capital circuit, so the rest of the urban hierarchy is being reshuffled into a series of key international 'regional' centres such as Los Angeles, San Francisco, Hong Kong and Sydney which control and service different parts of the Pacific Rim. In the process, the old, 'metropolis and region' structure of the early part of the twentieth century is in decline. Traditional national level cities will become international centres or regional cities. In many cities that were at the core of modern industrial capitalism, the new international division of labour has brought rapid de-industrialization. As Cohen (1981, p. 303) argues: 'changes in the international "competitiveness" of a number of world industries have been a major factor in the reshaping of the urban hierarchy throughout the world'.

Employment restructuring in London

London has played a key role in the international division of labour and urban hierarchy for well over a hundred years. During the nineteenth century London was the control and command centre of British global imperialism. As Dyos (1967, p. 28) put it:

> London had already become, not merely the country's capital, but the centre for most of the big commodity markets and financial houses, the country's chief port and industrial area. This commercial metropolis was being made more impregnable still by the concentrative power of the railways at home and the gains of economic imperialism abroad.

Charles Booth, the late nineteenth-century reformer, was grandiloquent and imperialistic:

> It is not only an unrivalled national emporium and world-market, but is also the Mother-city of the Kingdom and the Empire. London is at the centre, moreover, not only of the Imperial Government and of the Judiciary, but also of banking and finance, both national and international. It is in London that the agents-general for the Great Colonies, as well as the chief business agencies and official commercial representatives of foreign countries are to be found, their presence illuminating the fact that it is the recognized national centre, not only of government but of trade . . . Everything can be bought in London and therefore everyone comes to buy . . . London is as much an emporium for raw materials from all over the

world, as for finished products. (quoted in Steadman Jones, 1971, p. 161)

Looking at London's role in the national and international division of labour, it is clear that London's place as a centre of corporate and financial control and associated services has been strengthened and reinforced since the mid-1960s while its manufacturing importance has declined. In 1961 there were almost 1.5 million people working in London's factories. Since the mid-1960s, however, London has suffered rapid de-industrialization and by 1983 manufacturing employment had fallen to 600,000. Manufacturing employment declined nationally by 25 per cent between 1971 and 1981, but the decline was steeper and more widespread in London. London lost 40 per cent of its manufacturing jobs, and inner London lost 46 per cent between 1971 and 1981 (Hodgson, 1984; Serplan, 1988).

At the same time, the dominance of London as the location of the headquarters of large national and multinational companies and related producer services has been intensified. Westaway (1974a) found that 532 of the largest 1,000 British industrial companies and 86 of the top 100 companies had their head offices in London in 1971. The numbers increased slightly between 1969/70 and 1971/72, while other large cities lost head offices (Evans, 1973). A subsequent study by Goddard and Smith (1978) examining changes in corporate control in the British urban system between 1972 and 1977 confirmed this tendency toward greater centralization of head office functions in London. The rest of the South-east also showed an increase in the number of headquarters while the number decreased in every other region of Britain. This concentration of company headquarters in London during the 1970s was paralleled by a rapid increase in the number of offices of major overseas companies and of foreign banks operating in the City of London to take advantage of the London based Euro-dollar market which grew rapidly in the 1970s (Coakely and Harris, 1983). Dunning and Norman's (1983) examination of the location of some 261 US-based business services operating in the UK in 1976 found that 82 per cent of these operated from Central London, a further 13 per cent from locations elsewhere in London and only 5 per cent from places elsewhere in the country, mostly in the South-east. They also found that in 1976, 69 per cent of the local headquarters of foreign multinational manufacturing enterprises were located in London.

These trends are not new. Evidence from the first half of the 1960s pointed to the growing importance of financial services in London. In a comparative study of the British conurbation centres, Cameron and Evans (1973) showed that employment in professional and scientific services, insurance, banking and finance in London between 1961 and 1966 increased by three times the combined total for the other conurbations. This concentration of producer service employment has been subsequently reinforced. Daniels (1985) has shown that London and the South-east accounted for 50 per cent of the national increase in financial and professional services from 1978 to 1981. Greater London accounted for 20 per cent of all service jobs in Britain in 1981, but no less than 35 per cent of banking and finance employment. The GLC London Industrial Strategy stated:

> London is the capital of capitals: the head which directs the hand of labour on a national and international scale. As a trading centre, subsequently as a centre of finance and a market for services as well as a centre of government, London has always been a national centre of information and control . . . The concentration of control in London is also reflected in the fact that London has some 40 per cent of national commercial office space and that office rentals in central London reach their highest peak. (p. 334)

It would be a mistake, however, to conclude from these studies that white-collar employment as a whole has been growing in London. On the contrary, Daniels (1977) showed that while administrative, managerial and professional employment grew rapidly in London between 1966 and 1971, London was losing clerical workers (though more slowly than other conurbations). These trends have continued and they reflect the tendency towards the concentration of higher level decision-making and professional jobs in the information-rich, but highly expensive, central area and the decentralization of routine clerical and administrative operations which do not need an expensive central location to cheaper locations elsewhere in the South-east. This may account for the fact that employment in producer services grew far less rapidly in proportionate terms in London than in other regions and in the provincial cities over the period of 1971–81 (Hall, 1987). The growth in London was also far less than in the rest of the South-east. But, despite the marked decentralization of producer service employment, absolute growth of producer service employment in inner London between 1971 and 1981 was almost as great as the growth in all the major provincial cities put together. There is also evidence that R and D activities are concentrated in the South-east outside London (Howells, 1984). As a result, we are seeing the emergence of a spatially discontinuous division of labour: control and decision making are concentrated in London, R and D in the South-east and routine data processing and administration and production elsewhere (Massey, 1984).

Changes in employment and occupational structure in global cities

The rise of the new international division of labour and the switch from manufacturing to services is changing the spatial structure of employment and occupations as different industries have different spatial employment structures (Allen, 1988b). The growth of the financial and business services sector in global cities is leading to an increase in the proportion of professional, managerial and technical groups while the decline of manufacturing industry has led to a decline in skilled manufacturing jobs. In addition, Sassen-Koob (1984, 1988) argues that changes in the labour process and the spatial division of labour are generating a polarization of the occupational structure in global cities such as New York and Los Angeles as skilled manufacturing jobs are lost, and as the growing number of well paid professional and managerial workers is paralleled by a growing number of low-paid service workers in restaurants, hotels, cleaning, security guards, porters and other service jobs. She argues that there has been 'expansion in the supply of very high-income jobs, shrinking of middle-income blue- and white-collar jobs . . . and a massive expansion in the supply of low-wage jobs' (1984, p. 147).

This argument has been reinforced by Stanback and Noyelle (1982):

> For services as a whole there tends to be heavy concentrations of employment in better than average and poorer than average jobs. In contrast, in manufacturing and construction the distributions are more heavily weighted toward medium and average income jobs.

The thesis is clear. As manufacturing employment declines, and is replaced by service employment, occupational polarization can be expected to occur, with the growth of both highly skilled and lower skilled service jobs. As Ascherson (1986) has suggested:

> Under the surface of London, there are strange tremors. Something is rising to the surface, and when it bursts through it will create a new and almost unrecognizable type of urban society . . . that something is middle class money. London is ceasing to be a city with a large, organized factory proletariat

> and is reverting towards a capital in which a huge, underemployed proletariat works to service a wealthy minority.

The next section considers the extent to which this has occurred in London.

The changing occupational structure of London and the South-east

The concentration of corporate control and command functions and associated advanced services in London have long been reflected in the occupational structure of London and its commuting area. Westaway (1974b) analysed the changing occupational structure of economically active males in Metropolitan Economic Labour Areas (MELAs) in Britain between 1961 and 1966. Given the different occupational structures of men and women, such an analysis has a distinct gender bias, but Westaway showed that there were major differences in male occupational structures of different cities in the urban hierarchy. Professional and managerial occupations were over-represented in London and under-represented in the older industrial regions. Clerical jobs were also over-represented in the London employment area to a lesser extent. Conversely, the concentration of manual jobs was greatest in the industrial regions and lowest in London and the South-east. A more limited analysis by Waugh (1969) of the distribution of professional and managerial males in England and Wales in 1961 revealed a similar concentration of these groups in the South-east, particularly in the commuting areas outside London.

These findings are not surprising. The key question is the extent to which these occupational distributions have been reinforced as a result of the changes in industrial structure and the growing concentration of control and command functions in London and the South-east in the thirty years since 1961. Early work by Westaway found that between 1961 and 1966 there was a growth of professional and managerial jobs in the London MELA, a decline in the number of administrative clerical jobs and a major loss of manual jobs. Waugh also found a major growth of professional and managerial manpower in the areas around London concluding that: 'there is no parallel elsewhere in England and Wales to the rapid rise in status occurring within a ring 10–40 miles out from central London' (p. 166). These findings are very suggestive but they are now twenty years old. Significantly, more recent work by Green and Owen (1985) on the changing spatial distribution of socio-economic groups in manufacturing employment by labour market areas between 1971 and 1981 in Great Britain, confirmed the existence of marked differences in occupational composition and an increase in the proportion of employers and managers in the South-east outside London. They argue that the 'spatial dichotomisation' of the work force is being reinforced by the trends of 'professionalization' and 'proletarianization' in different parts of the country (p. 391).

More detailed work on the changing occupational distribution of economically active males and household heads in London and the South-east in the 1960s and 1970s (Hamnett, 1986, 1987) confirms the growing professionalization of the region. It also confirmed that both the absolute and the relative increases in professional and managerial groups were greater outside London than they were within. Residential decentralization of this group has taken place on a large scale. But although London lost skilled manual workers at a rapid rate over the twenty years, it lost semi-skilled and unskilled workers at an even more rapid rate. The end result has been that the socio-economic structure of London's economically active male population has *not* become increasingly polarized. Although the middle group of skilled manual workers has contracted sharply, the numbers and proportions of semi-skilled and unskilled workers have also fallen. London's occupational class structure has changed but it has become increasingly professionalized and low skill service employment has not absolutely expanded as some theorists and commentators have suggested. The only polarization has been in relative terms (Hamnett and Randolph, 1988).

Analysis of the changing occupational structure of economically active and retired women resident in London over the period 1971–81 revealed very similar results. There was a sharp increase in the number of women in professional and managerial occupations (+94 per cent), and the number of women in the intermediate non-manual category (teachers, nurses, technicians, inspectors, social workers, journalists, etc.) grew by 20 per cent. Simultaneously, the number of women in skilled manual occupations fell by 43 per cent, semi-skilled and unskilled occupations fell by 21 and 27 per cent respectively, and the number of junior white-collar workers (clerks, typists, telephonists, shop assistants, etc.) fell by 11 per cent. Overall, there was a sharp increase in the proportion of women working in the higher managerial and professional and intermediate non-manual groups and a decline in the proportion working in the manual categories. There was no evidence of a polarization in the occupational structure. The same was found where occupational change in inner and outer London was analysed separately. Nor did analysis of a 1 per cent longitudinal sample of London's enumerated population between 1971 and

1981 reveal evidence of a crude occupational polarization (Hamnett and Randolph, 1988). On the contrary, the growth of the service sector was strongly associated with a growing professional and managerial workforce. Low status service employment has not grown at a rapid rate. Instead, it appears to have been rapidly eroded. Occupational polarization in London has been very much a relative rather than an absolute phenomenon. Sassen-Koob's thesis may hold true in New York, but in London both the skilled manual and semi-skilled and unskilled groups rapidly contracted during the 1960s and 1970s. There was a slight relative polarization, but not an absolute polarization. Indeed, Glass (1973) has suggested an entirely different future for London from that derived from Sassen-Koob's arguments. Glass comments that:

> London is now being 'renewed' at a rapid pace – but not on the model about which we are so often warned. Inner London is not being 'Americanised': it is not on the way to becoming mainly a working-class city, a 'polarised' city, or a vast ghetto for a black proletariat. The real risk for inner London is that it might well be gentrified with a vengeance, and be almost exclusively reserved for selected higher-class strata. (p. 426)

Glass's fears have not been empirically demonstrated, but there has been a general upwards shift in socio-occupational composition with a rapid decrease in the numbers of skilled, semi-skilled and unskilled manual workers in London.

The changing income structure of London

Although the occupational structure of London has seen a tendency towards professionalization rather than polarization, the semi-skilled and unskilled still comprise an important part of the labour force, and there has been a trend towards a growing division in the income structure of those living in London. This tendency has been particularly marked over the last ten years as the recession threw large numbers of Londoners out of work and on to unemployment and supplementary benefit during the early 1980s. In the years 1979 to 1982 registered unemployment in London rose almost threefold from 132,000 to 400,000. The incidence of unemployment is strongly occupationally related, with unemployment rates being much higher among the less skilled. Second, the economic growth of the mid/late 1980s was associated with rapidly rising incomes for those employed in professional and managerial jobs in finance, banking and related producer service jobs: the growth areas of the Thatcherite entrepreneurial economy. This was particularly true in the City of London as the rapid expansion of jobs associated with the de-regulation of the City in October 1986 ('Big Bang') led to a bidding up of salaries for those with relevant skills. Thrift and Leyshon (1987) show that in 1987 the average income in the City of London was £20,000 – almost double the national average. But this figure includes large numbers of secretaries and low paid service workers – the incomes of the bankers, brokers, analysts and dealers are far higher.

While polarization in income distribution has taken place over much of Britain since 1979 (Halsey, 1987; Pond, 1989), it has been particularly marked in London. Research by Townsend (1987) shows that between 1979 and 1985 the lowest decile and the lower quartile of household incomes in Britain declined by 10 per cent and 7 per cent respectively in constant 1985 prices. Meanwhile the highest decile and upper quartile of household incomes increased by 10 per cent and 12 per cent respectively. But in Greater London, the highest decile and upper quartile of household incomes grew by 18 per cent and 12 per cent respectively, while the lowest decile and lower quartile of incomes fell by 17 per cent and 8 per cent respectively. In other words, incomes of the higher income groups rose by more in London than in Britain as a whole and the incomes of the lower income groups fell by more in London than in Britain as a whole (see Townsend, 1987, table 4.3). Thus, although occupational polarization has not occurred in London, it is clear that the income distribution has become more unequal, with higher income groups increasing their share at the expense of low income groups, whose incomes have fallen in real terms during the 1980s. In the next section I look at the links to housing market change.

The restructuring of the housing market and residential space in London

The change in the economic and occupational structure of London has been associated with a radical transformation of the tenure, price and spatial structure of the housing market in London, particularly within inner London. In 1961, London was still dominated by private renting. Almost two-thirds (64 per cent) of all households rented privately compared to just 19 per cent of households who were council tenants and the 17 per cent of owners. In outer London, by comparison, owner occupation was already dominant with 53 per cent of households and private renting was much less important (30 per cent). By 1981 the tenure structure of inner

London had been dramatically transformed. Private renting had fallen sharply to the point where it accommodated 30 per cent of households. Owner occupation had grown to 27 per cent of households and council renting to 43 per cent to become the largest single tenure in inner London. (Hamnett and Randolph, 1983; Hamnett, 1986b). So, in place of a dominant private rented tenure, council housing became the largest single tenure in inner London with a small but growing owner-occupied sector. Outer London, meanwhile, became more dominated by owner occupation.

This transformation has not been uniform across inner London. On the contrary, it has been associated with a geographical polarization of housing tenures within inner London. This polarization has occurred at a number of different spatial scales. First, sharp differences have opened up in the tenure structure of individual neighbourhoods as privately rented housing has given way to council redevelopment on one side of the street and high income owner-occupied gentrification on the other. Second, the long standing differences in the housing circumstances of the northern and western boroughs on the one hand and the eastern and southern boroughs on the other has been reinforced in terms of housing tenure. While the decline of private renting in the west has been accompanied by the growth of owner occupation and council renting, the growth in the east and south has been primarily of council renting. As a result, well over 50 per cent of households in some eastern boroughs are now council tenants. In Tower Hamlets in 1981 the proportion topped 80 per cent, while owner occupation accounted for just 5 per cent of households. The long standing differences between inner London and outer London and the rest of the South-east have also been perpetuated. While the dominance of owner occupation outside inner London has been intensified at the expense of private renting, the increase in the proportion of council renting was limited to inner London (Hamnett and Randolph, 1986b).

The change in housing tenures in London during the 1960s and 1970s accompanied the changes in economic and occupational structures. The term 'accompanied' is important, because it is *not* being argued that the changes in London's housing tenure were a result of economic and occupational changes. The two were largely autonomous and came about as a result of quite different sets of processes. The decline of private renting came about largely because renting had become unprofitable compared to the potential returns on sale to owners and the redevelopment of poor quality rented housing for council housing took place for social and political reasons. What *is* being argued is that the occupational restructuring and the restructuring of housing tenure were contingently related to changes in occupational structure through *household* income inequalities and the supply and demand of private housing and the price mechanism (see Chapter 1 on the key role of the household).

There are strong links between changes in occupational and income structure, and changes in the position of different groups within the housing market. In market economies, private housing is mainly allocated through the price mechanism which, in turn serves to balance supply and demand for housing. House prices in London have long been higher than the rest of Britain, partly as a result of the limited supply of owner-occupied housing, and partly as a result of the concentration of higher paid occupational groups who have bid up the supply of available housing. But, since the first house price boom in the early 1970s, London house prices have risen over 1,500 per cent to an average of about £100,000. In most of inner London, it is almost impossible to buy a house for under £150,000, and small one-bedroom converted flats fetch over £60,000. In parts of central London luxury houses can fetch over £1 million and large tracts of Docklands have been converted into high price flats for City workers. In practice, of course, what this means is that those with adequate income are able to buy, while those with low incomes are relegated to the council-rented sector, enforced sharing, overcrowding, short-life housing, bed and breakfast hotels or homelessness. A recent major study of homelessness in London (Bramley, *et al.* 1988) showed that the annual number of homeless acceptances by local authorities doubled from 1979 to 1986, and the number of households placed in bed and breakfast hotels increased ninefold from 1981 to 1987. The number of households officially accepted as homeless in 1986 was 30,000. In terms of the number of persons this is about 75,000 or 1 in every 100 Londoners. The number of single homeless persons sleeping rough, living in hostels or short-life accommodation or squatting is about the same again.

The housing market functions as a 'social sieve' (Glass, 1973, p. 421) or what Ball (1983, p. 287) has termed a 'spatial price mechanism', which sifts and sorts different income groups into different tenures, house types and areas according to their ability to pay. As Bell (1968) notes: 'the prices of houses act as a social sieve, with graduated meshes . . . through which drop each homogeneous segment of the population.' And Harrison (1983) states:

> British cities are segregated by class just as surely as British factories. No laws of apartheid are needed to enforce this separation: it occurs as naturally [sic] as oil divides itself from water, by the play of unequal incomes and savings in the housing market. (p. 180)

Harrison is correct as far as he goes, but the form and outcomes of the housing market sieving process at any time are influenced by both the available supply of housing and by the changing number, income structure and preferences of different households. As a result, changes in either demand for, or the supply of, housing of a given type, location and quality are likely to lead to changes in housing allocation among different occupational and income groups. Changes in demand are sometimes treated as though they are solely a result of changes in choice and preference, but changes in the structure of demand can and do occur as a result of changes in the occupational and income structure.

A variety of possibilities can be identified. A partial collapse of effective demand as a result of out-migration of higher income groups may lead, as it has in a number of old American inner city areas, to housing abandonment and dereliction. In these areas, the economically marginalized residual population, often black as well as poor, is left high and dry and effective demand moves out to the affluent suburbs with mortgage finance. Conversely, the growth of a highly paid professional and managerial labour force, is likely to lead to gentrification and place a considerable strain on the housing market as the price of desirable housing is rapidly bid up. Such shifts can lead to a re-assessment of previously ignored or run down housing areas and to the displacement or replacement of low income groups by higher income groups. It is also possible that a polarization of the employment, occupational and income structure can lead to a polarization of the housing market with both gentrification and abandonment occurring simultaneously in different parts of a city. Marcuse (1986) argues that this is occurring in New York where labour market polarization is linked to both gentrification and abandonment:

> Abandonment and gentrification are both reflections of a single long-term process, resulting from the changing economy of the central city. This process has two aspects: the shift from manufacturing to services, from reliance on mid-level skills to automation and de-skilling, on the one hand, which renders redundant large parts of the work-force and reduces lower-income rent paying ability; and the increasing professionalisation and concentration of management and technical functions, on the other, which creates additional higher-income demand for housing. These processes have spatial consequences: blue-collar workers (and potential blue-collar workers) are no longer needed in such large numbers down-town; professional and technical workers are in ever-increasing demand there. Housing adjacent to

> the central business districts reflects these changes. The pull exerted on one group by the changing economy of the CBD fits in with the push against the other. Thus the increasing polarisation of the economy is reflected in the increasing polarisation of neighbourhoods: at the one end, abandonment, at the other, gentrification. (pp. 154–5)

The gentrification and residualization of Inner London

The extent to which a gentrification of the inner city private housing stock takes place is dependent on the supply, price and location of available housing, on the level and nature of demand, and on the availability of mortgage finance (Williams, 1978). If, as has happened in many global cities, occupational and income shifts are associated with the growth of a young, highly educated professional and managerial class who work long hours and want to live as near to the city centre as possible, this is very likely to lead to gentrification, the existence of which has been well documented in London (Hamnett, 1973, 1976; Hamnett and Williams, 1980; Munt, 1987). Increased competition forces up prices and rents and squeezes out lower income groups who can no longer afford to live in the city. This is precisely what has happened in inner London. There has been a social and geographical expansion of the high income sector. As Glass (1973) pointed out at an early date:

> the 'colonising' drive of higher classes in Inner London has been accelerated: and so the working class quarters are becoming more constricted. Apart from the very rich, it is mainly the young members (or aspiring members) of the middle-upper strata, single people or couples without family responsibilities, who are prepared to pay the exorbitant housing prices of the inner sector. As these people live mainly in one- or two-person households, they have a disproportionately large number of households, and a disproportionately large housing demand, in relation to the total population size of their group . . . There are few hurdles in their path. The working class population of Inner London, old timers or recent recruits, are in every respect 'less equal' – not only in terms of income, status, expectations, but also in terms of age and dependency ratios. They have a higher share of old people, young children, large households. They find it difficult to resist being displaced or hemmed in. (pp. 425–6)

The changes which Glass identified have subsequently grown in importance. There has been an increase in the number of professional and managerial workers in parts of Inner London, particularly in the inner northern and western boroughs where gentrification has been proceeding rapidly (Hamnett, 1976, 1984, 1986, 1987; Hamnett and Williams, 1980; Munt, 1987). In Inner London as a whole, however, the increase has been very small in absolute if not in relative terms, largely because of the large absolute fall in the number of usually resident professional and managerial households in the three central boroughs of Camden, Kensington and Chelsea and Westminster. The fall in central London has been partly associated with the decline of the multi-occupied privately rented sector, and the sale of flats for owner occupation (Hamnett and Randolph, 1988) which have both been associated with a decline in the number of what the census term 'usually resident household spaces'. In other words, residents in central London are being squeezed out by reductions in the number of places to live as a result of the conversion of multi-occupied private rented households for single family owner occupation.

This process of tenure transformation was associated with a process of socio-tenurial polarization whereby the higher social classes who were recolonizing the inner city moved predominantly into owner-occupied housing or converted previously private rented multi-occupied housing into single-family housing. At the same time, the less skilled, the lower paid and various minority groups who had previously lived in poor quality private rented housing became increasingly concentrated in the expanding council sector (Hamnett, 1984; Bentham, 1986; Hamnett and Randolph, 1987). The long term consequences of this process were foreseen by Tindall (1973):

> The turn of the century seems likely to see the whole of inner London parcelled out either into council blocks . . . or into highly expensive and inevitably exclusive owner-occupied property. The gain in overall appearance and repair may be considerable, the loss in flexibility and in the possibility of living space for those who are neither well off nor clever, will be immeasurable.

And as has been indicated, this social polarization of the Inner London housing market has been associated with the geographically uneven form of the tenure transformation. The geographical distribution of tenures and the geography of occupational class have become inextricably associated in the social and tenurial polarization of residential space. The professional and managerial gentrifiers – the 'space invaders' as Smith and Williams have termed them – have colonized or recolonized an increasingly large part of inner London, leaving the economically marginalized and residualized manual working class increasingly concentrated in the inner city council sector – the only housing to which they can gain access. Meanwhile, growing and equally unaffordable owner occupation in outer London has thrown a 'cordon sanitaire' around the inner London working class. The working class were driven out of central London in the late nineteenth century. In the late twentieth century they have been driven out of large parts of the inner city and forced back into council reservations. This has led to increasingly sharp residential divisions between the haves and have nots. As the GLC London Industrial Strategy (1985) put it:

> There is no clearer example of the distance between them than in Tower Hamlets. There, the council blocks stand facing the commercial skyscrapers of the City: the one representing one of the most depressed areas in Britain, the other sited in one of the richest square miles in Britain. It is a contrast that has been seen before in London's past. (p. 8)

But while the GLC are correct to point to the distance between the haves and have nots and between the capital of capitals and the barrack blocks of the less skilled, low paid and unemployed, the two phenomena are not unrelated. They are part and parcel of a wider process of economic and social restructuring which has created a large and growing professional and managerial service class and left the rump of the less skilled washed up on the sands of the inner city – to use Malcolm Cross's graphic image. Although other countries and other parts of Britain possess marked social and spatial contrasts, such contrasts are particularly acute and visible in the global cities where the affluent and the very poor live within close proximity of one another. In the East End, the derelict and abandoned docks have been transformed in just five years from an urban black hole into a yuppie paradise and property market hot spot. The same is happening a mile or so north of the City in Islington and Hackney where decaying Georgian and Victorian houses have been transformed into luxury homes. Yet, only a few hundred yards away from these areas are major concentrations of poverty, unemployment and poor housing in the decaying council estates.

Conclusions

It has been argued that changes in the organization of production and the social and spatial division of labour have led to the restructuring of the urban hierarchy and to the emergence, or more accurately, the growing dominance of a small number of global cities. By virtue of the concentration of corporate head offices and associated financial and producer services in these cities they play a key role in the control and co-ordination of production and exchange at the international level. It was also argued that this concentration of high-level control and servicing functions had led to the growing professionalization of the occupational structure of these cities. They are one of the main concentrations of the service class in contemporary society. It has also been argued by some commentators, though the empirical evidence for this is lacking in London, that such cities have seen the absolute growth of a low-skill 'servicing class'. When this is combined with the decline of manufacturing industry and the reduction in the number of skilled manual workers, the result, it is argued, is a polarization of the occupational structure between the service class and a servicing class. This does not appear to have occurred in London except in a limited relative fashion. Instead, London has seen a large relative expansion of the service class and a shrinkage of all other occupational groups. London does, however, appear to have seen the growth of income polarization although more detailed empirical work is necessary in this area.

Following Randolph in Chapter 1, it was also argued that the occupational and income structure of different areas interacts, through the mediating role of households, with the supply, price and tenure structure of the housing market to produce a socially and spatially segmented housing market. The form of this segmentation is contingent on the structure of the housing market. In London, however, the decline of the private rented sector and the growth of home ownership and the council rented sector have meant that rising demand for home ownership and rapidly rising prices have led to the concentration of the more highly skilled and highly paid service class in this sector and the concentration of the less skilled and low paid in the council sector. The result has been growing occupational and income polarization by tenure. Thus, changes in labour market structure have interacted with changes in the tenure structure of the housing market to produce a socially and spatially segmented residential structure in London.

References

Allen, J. (1988a), 'Service industries: uneven development and uneven knowledge', *Area*, 20, pp. 15–22.

Allen, J. (1988b), The Geographies of Service, in D. Massey and J. Allen, (eds) *Uneven Re-development* (Sevenoaks, Hodder and Stoughton).

Ascherson, N. (1986), 'London's New Class: the Great Cash In', *The Observer*, 25th May 1986.

Ball, M. (1983), Housing Provision and the Economic Crisis, *Capital and Class*, 17, 60–7.

Bell, C. (1968), *Middle Class Families* (London, Routledge and Kegan Paul).

Bentham, G. (1986), Socio-tenurial polarisation in the United Kingdom, 1953–83: the income evidence, *Urban Studies*, vol 23 pp. 57–62.

Bramley, G. *et al* (1988), *Homelessness and the London housing market* (Bristol, School for Advanced Urban Studies).

Coakely, J. and Harris, L. (1983), *The City of Capital* (Oxford, Basil Blackwell).

Cameron, and Evans, A. (1973), *Regional Studies*, vol 7, pp. 47–55.

Cohen, R. B. (1981), 'The new international division of labour, multinational corporations and urban hierarchy' in M. Dear and A. J. Scott, (eds) *Urbanization and Urban Planning in Capitalist Society* (London, Methuen).

Daniels, P. (1977), 'Office Location in British Conurbations: trends and strategies' *Urban Studies*, 14, pp. 261–74.

Daniels, P. (1985), 'Producer services and the post-industrial space economy' in R. Martin and B. Rowthorne (eds) *The Geography of De-Industrialisation*, (Basingstoke, Macmillan).

Dunning and Norman (1983), 'The theory of the multi-locational enterprise: an application to office location', *Environmental Planning a*, 15, pp. 675–92.

Dyos, J. (1967), 'The Slums of Victorian London', *Victorian Studies* September, pp. 5–40.

Economist The, 9 September 1989, Tokyo.

Evans, A. W. (1973), 'The location of the headquarters of the industrial companies', *Urban Studies*, 10, pp. 387–95.

Goddard, J. B. and Smith, I. J. (1978), 'Changes in corporate control in the British urban system, 1972–77' *Environment and Planning A*, 10, pp. 1073–84.

Glass, R. (1973), 'The Mood of London', in *London, Urban Patterns, Problems and Policies*, D. Donnison and D. Eversely (eds), Heinemann.

Greater London Council (1985), *The London Industrial Strategy*.

Green A. and Owen, D. (1985), 'The Changing Spatial Distribution of

Socio-Economic groups employed in manufacturing in Great Britain, 1971–1981' *Geoforum*, 16, 4, pp. 387–402.

Hall, P. (1966), *The World Cities* (London, Weidenfeld).

Hall, P. (1989), 'The anatomy of job creation', *Regional Studies*, 21, pp. 85–106.

Halsey, A. H. (1987), 'Social Trends since World War II' *Social Trends*, 17, pp. 11–19.

Hamnett, C. (1973), Improvement Grants as an Indicator of Gentrification in London, *Area*, 1973, 5, 4, pp. 252–61.

Hamnett, C. (1976), Social change and social segregation in inner London, 1961–71, *Urban Studies*, 13, 3, pp. 261–71.

Hamnett, C. (1983), 'Split City', *Roof*, July/August pp. 13–14.

Hamnett, C. (1984), Housing the Two Nations: Socio-Tenurial Polarisation in England and Wales, 1961–81, *Urban Studies*, 21, 3, pp. 389–405.

Hamnett, C. (1986a), The changing Socio-Economic Structure of London and the South East, 1961–81, *Regional Studies*, 20, pp. 391–406.

Hamnett, C. (1987), A Tale of Two Cities: socio-tenurial polarisation in London and the South East, 1966–81, *Environment and Planning A* 19, pp. 537–56.

Hamnett, C. and Randolph, W. (1987), 'The residualization of council housing in Inner London', in D. Clapham and J. English (eds.) *Public Housing: Current Trends and Future Developments* (London, Croom Helm).

Hamnett, C. and Randolph, W. (1983), The changing tenure structure of the Greater London Housing market 1961–81, *The London Journal*, 2, pp. 153–64.

Hamnett, C. and Randolph, W. (1988), Labour and Housing Market Change in London: A Longitudinal analysis, 1971–81, *Urban Studies* 25, pp. 380–98.

Hamnett, C. and Williams, P. (1980), Social Change in London: a study of gentrification, *Urban Affairs Quarterly*, 15, pp. 469–87.

Harrison, P. (1983), *Inside the Inner City*, (Harmondsworth, Penguin).

Hodgson, M. (1984), *Economic Statistics from the 1981 Census*, Statistical Series, no 34, Greater London Council.

Howells, J. R. L. (1984), 'The location of research and development: Some observations and evidence from Britain'. *Regional Studies*, 18, pp. 13–19.

Marcuse, P. (1986), Abandonment, gentrification, and displacement: the linkages in New York City in N. Smith and P. Williams, (eds) *Gentrification of the City*, (London, Allen and Unwin).

Massey, D. (1978), 'In What Sense a Regional Problem?' *Regional Studies* 13, pp. 233–43.

Massey, D. (1984), *Spatial Divisions of Labour*, (Basingstoke, Macmillan).

Munt, I. (1987), Economic restructuring, culture and gentrification: a case study of Battersea, London, *Environment and Planning, A*, 19, pp. 1175–97.

Pond, C. (1989), 'The changing distribution of income, wealth and poverty', in Hamnett, McDowell and Sarre, (eds) *The Changing Social Structure*, (London, Sage).

Sassen, S. (1988), *The Mobility of Labor and Capital* (Cambridge, CUP).

Sassen-Koob, S. (1984), 'The new Labour Demand in Global Cities' in P. Smith (ed.) *Cities in Transformation*, Urban Affairs Annual Reviews, vol. 26, (Beverley Hills, Sage).

SERPLAN, (1988), *Regional Trends in the South East*, RPC 1060.

Smith, N. and Williams, P. (1986), (eds) *Gentrification of the City*, (London, Allen and Unwin).

Soja, E. (1986), 'Taking Los Angeles Apart', *Society and Space, 4*, pp. 255–272.

Stanback, T. M. and Noyelle, T. (1982), *Cities in Transition*, (Totowa, N.J. Rowman and Allenheld).

Steadman-Jones, G. (1971), *Outcast London* (Oxford, OUP).

Tindall, G. (1973), 'A Street in London', *New Society* January 14th.

Thrift, N., Daniels, P. and Leysham, A. (1989), 'Sexy, greedy: the new international financial system', *Working Papers on Product Services, 8*, University of Bristol.

Townsend, P. (1987), *Poverty and Labour in London*, Low Pay Unit.

Waugh, M. (1969), 'The changing distribution of professional and managerial manpower in England and Wales between 1961 and 1966', *Regional Studies*, 3, pp. 157–69.

Westaway, J. (1974a), 'The spatial hierarchy of business organization and its implications for the British urban system', *Regional Studies*, 8, pp. 145–55.

Westaway, J. (1984b), 'Contact potential and the occupational structure of the British urban system', *Regional Studies*, 8, pp. 57–73.

Williams, P. (1988), 'Building societies and the inner city', *Transactions of the Institute of British Geography*, 2, pp. 23–34.

8 *Producing a divided society: state policies and the management of change in housing and labour markets in a peripheral region*

RAY HUDSON

Introduction

In this chapter I want to focus on how state policies have mediated various forms of change in and between labour and housing markets in North-east England (Fig. 8.1). Such changes can and do come about through a variety of mechanisms and institutions and the justification for focusing on the state in this context is a simple one, though it raises more complicated theoretical and political issues. It can be stated as follows: over the last four decades or so there is a very real sense in which the North-east has become a state-managed region. The deprivation, poverty and unemployment of the interwar years was to be abolished by a state-managed sectoral and spatial reorganization of labour markets and housing provision, with public sector construction and allocation policies mediating the play of housing market forces for most of the population. What actually happened is another and different story. After a prolonged period of considerable state involvement, the North-east is again characterized by high unemployment and widespread poverty. The heady optimism of the 1940s has evaporated and the 1980s look increasingly like the 1920s and 1930s for many people . . . precisely what increased state involvement was supposed to prevent. There have certainly been dramatic changes in housing and labour markets but there is a yawning chasm between actual and intended changes. The remainder of this chapter explores the dimensions of and reasons for this.

State policies and regional changes in a period of 'full employment', 1942–58

During this period, the over-riding priority shaping state policies and those of major capitalist interests in the North-east was guaranteeing stability in the market for male labour. Industries that had been deep in crisis in the interwar years were revived to profitability by the demands of the war economy and the subsequent period of postwar reconstruction. By 1942 unemployment had virtually been eliminated, ushering in a period of 'full employment'. Companies producing commodities such as coal, steel, ships and armaments were desperate to preserve their labour forces to ensure unfettered profitable production. The state was desperate to prevent labour shortages endangering output in what were now seen as vital branches of industry, above all in coalmining.

The UK economy was almost exclusively dependent upon coal as a primary energy source and there were great pressures to prioritize maximization of coal output. Nationalization in 1947 seemingly offered a mechanism through which the state could pursue this objective. Attaining it, however, proved problematic. Although increasing output via selective investment in some collieries remained the medium-term strategy for boosting output, in the short term increasing output depended upon maintaining the mining workforce and thus to steering alternative male-employing manufacturing activities away from the coalfields. Consequently, there was little wartime decentralization of production to the North-east. What was moved to the region was mainly to non-coalfield locations such as Newton Aycliffe, and temporarily drew married women into the wage labour force (Pepler and MacFarlane, 1949). Subsequently, the NCB, powerfully supported from the Ministry of Fuel and Power, successfully vetoed the introduction of new male-employing manufacturing plants into coalfield areas by the Board of Trade as part of the new, strengthened postwar regional policy. To do so would invite an exodus from the mines to better paid jobs in healthier, safer working environments. These arguments about the need to preserve the coalmining labour force and protect it from labour market competition were largely accepted by mining trades unions and local authorities on the coalfields. These authorities were Labour controlled and with substantial political representation of mining interests on them. To local authorities, coalmining seemed to offer a stable economic base for their areas. Trades union leaders saw such policies as offering long-term employment to *their* members and industrial relations were constructed to incorporate union leaders in participatory structures which enhanced

Figure 8.1 Main urban areas of the North-east of England

their status as part of an attempt to ensure strike-free production.

A considerable consensus emerged, then, committed to preventing changes in the male labour market on the coalfields, drawing support from key central government departments, the NCB, local authorities and mining trades unions. But such pressures were not confined to the coalfield areas. The war economy had revived demand for the North-east's 'traditional' manufactured goods: steel, ships, heavy engineering products, armaments and so on. Producing these again became profitable; postwar reconstruction, the demand for armaments generated by the Korean War and a climate of mild international competition continued the recovery in profits. Output boomed in these industries, based on fuller working of existing capacity. Consequently, as in coal, boosting output and profits crucially relied upon increasing and retaining a mainly male labour force. Similar pressures arose in the main manufacturing locations – the major urban areas of Teesside, Tyneside and Wearside – as on the coalfield to prevent the introduction of potential competitors in the male labour market. On Teesside these pressures were compounded by growing labour demands at ICI's massive new Wilton petro-chemicals complex. This was regarded as of national significance and state policies were shaped to ensure that labour shortages did not emerge there (Beynon, Hudson and Sadler, 1986b).

There were, then, formidable and widely supported pressures to freeze the pattern of economic activities around those industries which had collapsed in the interwar years and been instrumental in the emergence of the 'regional problem'. Moreover, their male labour forces were to be preserved, protected from competition on the labour market. In other ways, however, labour market changes were seen as both acceptable and desirable. Established employers of male labour did not object to the introduction of predominantly female-employing activities. Furthermore, such 'diversification' was seen as desirable by local authorities and the Board of Trade. Thus, especially between 1945 and 1947, there was some locationally selective expansion of capacity of mainly female employment in clothing and electrical and instrument engineering. The availability of female labour in the North-east coincided with pressures on the female labour market in regions such as Lancashire and Yorkshire. But the same pressures – expressed in the patriarchal culture of the 'heavy' industrial areas – that historically had resulted in little participation of married women in the wage labour force were being strengthened by the priority given to preserving the 'traditional' pattern of production and the associated demand for male labour, with its systems of shift working, especially in coal and steel. Nevertheless, female wage labour continued to expand slowly in clothing, electrical engineering and paper, printing and publishing during the 1950s, with output growth linked more to rising labour productivity than labour force expansion. This reflected the incorporation of 'green', though not necessarily non-unionized, labour into the regimes of factory work for the first time, allowing labour processes to be more easily reorganized and intensified, providing substantial productivity gains in activities where the emphasis often lay upon unskilled assembly work (North Tyneside Community Development Project, 1978). In fact, the greatest expansion of female wage labour was in retailing as increasing working-class purchasing power led to growth in this activity, and in education and health, linked to expansion of the welfare state, partly socializing the costs of reproducing labour power and, to a degree, decommodifying provision of such services. Growth in female wage labour, then, was both modest in scope and varied in cause but it was a harbinger of broader changes in patterns of employment, class and gender relations that were to follow.

Expansion in female wage labour was partly related to changes in housing markets and, more generally, living conditions for some sections of the working class. Part of the postwar settlement involved increasing working-class housing standards via an expanded public sector programme (Short, 1982). In the North-east pursuit of this objective became intertwined with that of achieving desired male labour market stability and selective female labour market change via the formation and implementation of local authority settlement policies. The origins of such policies lie in the interwar years when the view emerged that some places, especially former colliery villages, would never again be integrated into the mainstream of capitalist production and so had lost their economic rationale for existing. This viewpoint evolved over the intervening years to emerge after 1947 in local authority planning policies in more generalized form, categorizing settlements in terms of whether – and to what extent – they should receive public sector investment in housing and services. The criterion for deciding which settlements should become 'growth settlements' was their future economic prospects. Settlements with what were believed to be 'long-life' collieries, steelworks, shipyards and so on were those where local authorities, New Town Development Corporations and government departments concentrated their new infrastructive investment. Even so, implementing such policies to produce a spatial reorganization in service provision and the housing market was in some respects problematic. District councils as housing authorities often opposed county councils as planning authorities over designation of settlements in their areas as 'non-growth' ones. While central government via the Ministry of Housing and Local Government generally turned down their appeals and supported county council policies, the resultant delays prevented a smooth implementation of settlement policies and the intended spatial reorganization of public sector housing provision.

Concentrating 'better' public sector housing around allegedly 'long life' points of production and creating better living environments was seen as one way of helping stabilize male workforces and simultaneously providing some of the necessary pre-conditions to attract new female-employing industries. But growth of female wage labour could help resolve another conflict that was inherent in pursuing goals of improving working-class housing and living conditions while seeking to hold down wage increases in 'traditional' male employing industries. Women's wages – especially married women's wages – could help meet the increased costs of living associated with changing consumption patterns. These were perhaps most acutely – but by no means exclusively – felt in coalmining areas. Many miners lived in NCB housing, often of poor quality but for which they paid low rents. Provision of better quality housing and living environments had been an important element in the 1945 Miners' Charter (Horner, 1960, pp. 176–9). Moving to a new local authority or New Town Development Corporation house, however, meant a sharp jump in rents. This often provoked problems, especially given the pressures exerted by the government on the NCB to hold down miners' wages. In Peterlee, for example, a New Town built to provide better housing for miners, led to considerable conflict between the Development Corporation and tenants and to rent strikes in the 1950s (Robinson, 1978). This is indicative of the broader incompatibilities between the goals of increasing working-class living standards and the imperatives governing the region's 'old' industries, in both the public and private sectors.

In summary, this 1942–58 period was characterized by a broad consensus within the North-east on the desirability of maintaining the 'traditional' structure of industrial production and pattern of male labour demand associated with it. Diversification of economic activities and employment was restricted to those spheres where growth would be of wage labour for women. This imperative to freeze the existing pattern of economic activities and labour demand informed the formation and implementation of state housing and settlement policies within the region. In a sense, there was also a consensus – though not necessarily a conscious one – on the desirability or at least necessity of freezing existing patterns of class and gender relations within the North-east. However, changes in the housing market produced pressures that required some alterations in these existing relations and served both to redefine and produce divisions within the working class related to consumption patterns and lifestyles. Even so, there were limits to these changes. An increase in women working for a wage offered one way of meeting increased housing costs yet the reproduction of the 'traditional' pattern of production and male labour demand posed a formidable barrier to any significant increase in female wage labour. From 1958, however, this barrier began to dissolve and the thrust of state policies was directed, in various and often conflicting ways, towards producing generalized labour market changes.

State modernization policies to preserve 'full employment' and regional changes, 1958–75

State involvement in generating labour market change took one of two main forms. One strand of state policies was primarily directed towards diversifying employment opportunities, introducing new industries and employers of wage labour into the North-east. This concern to provide new jobs reflected the very visible direct involvement of the state in breaking up existing patterns of production and employment, especially in coal and steel after its nationalization in 1967, and less directly but equally significantly in shipbuilding (see Hudson, 1985). From 1958 previous pressures to freeze the region's economic structure were weakened because of changes in international markets for energy and ships (which in turn reduced demand for steel) and decisions by the state, as restructuring 'old' industries became central to the government's intentions to modernize manufacturing in the UK to restore international competitiveness (for an extended discussion, see Hudson, 1988).

Employment decline in shipbuilding and, especially, coalmining led to the recreation of a substantial locationally concentrated unemployed population within the North-east at a time of tight labour market conditions in the growth regions of the UK and other major capitalist states. For many capitals, availability of labour-power alone provided a compelling attraction, providing opportunities to recruit a variety of types of labour and, between 1967 and 1976, at a price directly reduced via the Regional Employment Premium. Some companies came in search of male labour, socialized into the disciplines of three-shift working in the mines. Others sought compliant 'green' labour, often female. Promises of trouble-free production in exchange for closed shop arrangements as trades unions competed for members were a further inducement.

The result of this enhanced flexibility and managerial control over labour was increased labour productivity in many branches of manufacturing (Northern Region Strategy Team, 1975).

In addition to this, the North-east offered a particularly attractive environment to manufacturing capital for other reasons, linked to working-class and state responses to employment decline and unemployment growth. Mounting job losses in 'old' industries were generally accepted as inevitable and went uncontested by trades unions, especially at their higher levels, partly because of a belief that central government's strengthened regional policy and new regional planning machinery would provide alternative employment. Within the North-east, the intention was to create new general conditions of production and environments designed to attract fresh manufacturing investment and employment in new industries into selected locations, such as New Towns. This was to be achieved via locationally selective increases in public expenditure, allied to new forms of planning. Such policies undoubtedly met with some initial success (Hudson *et al.*, 1976, vol. 1, pp. 139–40).

The North-east also became a more enticing environment for private capital because enhanced central government financial incentives became available through regional industrial policy. This made it an attractive destination for 'capital intensive' production plants, with a high technical composition of capital. These incentives were associated with diverse labour market changes. They did lead to some inward movement of new plants, mainly of US-based multinationals in chemicals and pharmaceuticals. Such new projects provided few jobs, however, and the main effect of financial incentives was to facilitate *in situ* restructuring and job shedding in existing chemical plants (Northern Region Strategy Team, 1976a; Hudson, 1982a). In this sense, implementing regional policy measures was the proximate cause of rising unemployment – diametrically opposite to the intended policy effect on the labour market.

More generally, the combined and interacting effects of state policies, corporate strategies and trades union tactics led to some diversification of employment opportunities by the early 1970s. To this extent, labour market changes were as intended. There was, however, minimal net growth in manufacturing employment. Gains in 'new' industries were offset by losses in 'old' ones. Moreover, total industrial employment fell sharply because of coalmining losses. Furthermore, diversification was linked to significant qualitative changes in the types of job available and in the gender composition of the workforce. The diversity of new industries was accompanied by a relative homogenization in terms of labour processes and types of work as the region became first a 'national outpost', then a 'global outpost' (Austrin and Beynon, 1979), for a time the location of branch plants that formed one link in increasingly global chains of corporate production (Northern Region Strategy Team, 1976b; North-east Trades Union Studies Information Unit, 1977).

Important changes were also occurring in the service sector of employment within the region, reflecting decisions about the level and organization of state involvement. The gender composition of the workforce in service activities was altered. Net employment for men fell (largely because of the decline of activities dependent upon coalmining) while that for women, especially married women, rose, though often on a part-time basis. There were also changes in the sort of paid work undertaken by women: decline in activities such as distribution was counter-balanced by growth in the welfare state and the relocation of routine operation within the administrative structure of the state itself. This considerably increased the influence of public service sectors unions within the region.

But perhaps the most significant aspect of labour market change in this period is the simultaneous growth of net out-migration from the region and unemployment within it. Even in the relatively favourable economic climate of the 1960s – one of strong economic growth by UK standards – the volume of 'new' jobs was insufficient to counter the losses of 'old' ones, with significant differences in the skill and gender composition of jobs lost and gained. Net male employment fell while female employment grew; half the new manufacturing jobs and most of the new 'service' jobs were filled by women, increasingly by married women. Certainly there was 'diversification' in employment opportunities, as state policies intended, but it was not associated with a return to 'full employment'. This disjunction reflected both changes in labour processes, the recruitment practices of employers and prevailing notions about relationships between the social and technical divisions of labour. Increasingly, there was a growing division between households with two (or more) wage earners and those with none and this was linked with the uneven transformation of consumption patterns and lifestyles within the region.

For the modernization policies of the 1960s were concerned with engineering much more all-embracing changes, of which those in production patterns and the labour market were only one part. A central element of these policies was a restructuring of housing provision, both locationally and in terms of tenure. The settlement classification policies which the county councils had introduced in the 1950s were endorsed and reinforced, with the

1963 Hailsham programme for regional modernization (discussed more fully in Hudson, 1988, chapter 3) providing the final push leading to a third New Town being designated under the 1946 Act at Washington. In addition, Northumberland County Council undertook to develop two new towns at Cramlington and Killingworth via the 1952 Town Development Act (Rowntree Research Unit, n.d.), strengthening the emphasis on intra-regional spatial selectivity in modernization policies. Nevertheless, opposition 'from below' to such policies continued as residents of villages designated as no-growth locations pressured district councils to challenge the rationale of county council settlement policies, delaying or even preventing the 'planned' changes in patterns of new housing provision. As well as new industry being steered to New Town locations, new public sector housing, often involving experimenting with cost-cutting system building techniques, was concentrated in them and other growth points. As its architect, Roy Gazzard, later admitted of Killingworth: 'We were building a town on the cheap . . .' (*Newcastle Journal*, 26 November 1985). But this was just one example of a proliferation of system building both to replace nineteenth-century housing in the riverside conurbations (for example, see Benwell Community Development Project, 1978b) as well as in 'growth points' such as Aycliffe, Peterlee, Spennymoor and Washington, often on a 'high rise' basis. The powers and resources of the New Town Development Corporations were undeniably crucial in enabling 'growth point' policies to be implemented successfully. They could undertake substantial capital investment in new housing and go into debt on their housing revenue accounts in the process. Over the period 1957–72, for example, for each net addition to the public sector stock in the remainder of County Durham, there was a net addition of two houses in the three Durham New Towns (Hudson *et al.*, 1976, vol. 1, pp. 192–4). This locational concentration of new public sector housing, allied to intra-regional migration and increased commuting associated with rising per capita car ownership, served to assemble pools of labour from which companies could select workforces for their new factories while housing allocation policies prioritized the needs of 'key workers' (Robinson, 1978; Hudson, 1982).

In addition, a growing prominence was given to encouraging private capital in construction to provide 'executive housing' and, more generally, speculatively built housing for owner occupation for some strata of the working class as well as the 'new middle class', largely created through expansion of state activities. Again, new towns played an important role. The development of Cramlington perhaps best exemplifies these tendencies, planned from the outset as a *de facto* private sector New Town with an 80 per cent owner-occupation rate. But increasingly throughout the 1960s greater emphasis was put upon owner occupation in New Towns designated under the 1946 Act. From the outset, Washington had a target of 33 per cent owner occupation. In 1967 this was raised to 50 per cent for all such New Towns as a result of a central government decision as increasing owner occupation became a shared goal in the national housing policies of both Conservative and Labour parties. Subsequently, the decision was taken to sell New Town houses to sitting tenants at a 20 per cent discount (see Cullingworth and Karn, 1968) to encourage further growth of owner occupation.

Such tenure changes, as well as provision of 'better' working-class public sector housing, were closely linked to labour market changes and the growth of married women working for a wage. The stated intention within regional modernization policies was to produce a parallel but selective transformation in consumption patterns and lifestyles, with the growing incorporation of married women into the wage labour force as the crucial link between changes in production and consumption. They would provide cheap, pliant labour for 'new' economic activities while their wages would complement those of their husbands to produce an increased household consumption fund. This would allow increased expenditures on housing – above all the switch into owner occupation – cars and consumer durables while ownership of labour saving devices such as fridges, freezers and washing machines would facilitate the growth of married women working outside the home (see Hudson, 1980a, 1980b). But the intention was also that these changes would be associated with a switch to more privatized lifestyles and changed attitudes towards industrial relations and politics. The growing emphasis upon the household centred on the nuclear family helped cut the ground from beneath notions of collective action and attachments to more broadly based social groups. Modernization of consumption thus produced further divisions within the working class.

This was especially so as spatially selective housing allocation policies were also devised to help 'manage' those more or less permanently unemployed and surplus to the labour-power requirements of both the state and private capital. Largely reliant upon state transfer payments for a minimal level of subsistence, such people were increasingly concentrated into 'sink' housing estates as a result of public sector allocation procedures (Taylor, 1979; Byrne and Parson, 1983). Spatially segregated in this way, in locations that could be more heavily policed and subject to increased surveillance by social security officials, this surplus population was more amenable to management by the state, safely out of sight and away from the two-income mass consumption lifestyles of other social groups. By selectively allocating public sector housing, channelling the unemployed and those categorized as 'social problems' (single parent families, the mentally ill and so on) into particular estates that then became stigmatized as 'problem areas', the unemployed could be removed from the sphere of everyday experiences of the employed and 'normal' population. Problems of poverty and ill-health associated with localized pockets of very high unemployment were safely confined to precisely delimited areas of public sector housing. In this sense, housing allocation policies depoliticized the problems of unemployment while at the same time living on one of these stigmatized estates militated against finding a job and being allocated to 'better' quality public sector housing. This was one of the more obvious manifestations of the divisiveness – between and within classes – that was an integral part of the modernization process as particular patterns of change in labour markets and housing provision were set in motion through state activities. For if, in contrast to the preceding period, the leitmotiv of state policies in this period was modernization and change to reduce inter-regional differences, the impress of modernization was intentionally to be experienced within the region in very different ways by different social groups and localities. This differentiation in terms of position in both labour and housing markets was partly achieved as intended but there was also emergent evidence of unintended effects. The state became increasingly incapable of managing and co-ordinating these changes so as to produce only those effects that it intended.

The abandonment of modernization policies: regional change in an era of disinflation policies, 1975–87 (and beyond . . .)

The most striking feature of labour market change in the North-east in this period is the tremendous expansion in unemployment, reminiscent of the 1920s and 1930s. Over much of the region, including many of the 1960s 'growth points', registered unemployment rates exceeded 20 per cent and, in smaller areas, 40 or 50 per cent, by the mid-1980s. As the Department of Trade and Industry (1986, para. 7A. 1.16) conceded: 'The North East is in the unenviable position of having the worst combination of unemployment blackspots in the country.' This aggregate growth reveals but half of the story, however, for the character of unemployment was also changing, increasingly affecting all age groups, men and women alike, with more people (over 50 per cent of those registered) becoming 'long term' unemployed (DTI, 1986, paras. 7A. 1.17 and 18), permanently surplus to requirements for labour-power. The proximate causes of this unemployment expansion lie in the combined effects of international recession (Mandel, 1978) and the policies of the state towards it. From 1975 the Labour Government adopted a policy of reducing and altering the composition of public expenditure; subsequently, the Conservative Government intensified this switch. It had particularly severe effects on the labour market in the North-east because of its impacts on nationalized industries, the welfare state and on central government regional policy expenditures.

The early 1970s seemingly promised a rosy future for the nationalized coal and steel industries in the North-east but major expansion plans were soon transformed into savage cut-backs (Hudson, 1985; Beynon, Hudson and Sadler, 1986). The nationalization of shipbuilding in 1977 was simply the prelude to rationalization (Hudson, 1985). Capacity closures and mass redundancies became the order of the day in shipbuilding and steel towns and pit villages, leading to soaring unemployment rates, both as a result of direct job losses and indirect ones as multiplier effects worked through to local service sector jobs (Hudson and Sadler, 1984; Hudson, Peck and Sadler, 1985). With the major exception of the 1984–5 miners' strike, these job losses were largely uncontested (though see Hudson and Sadler, 1983).

Employment growth in the North-east over the two decades prior to 1975 had been mainly linked to expansion of the welfare state and also in the administrative activities of the state. Planned decentralization of central government activities, notably the Property Services Agency, was scrapped while cuts were made in existing local and central government jobs. The rate of employment growth in education and health was slowed, then halted, as public expenditure cuts affected these services. Pressures to increase employment on a part-time basis, already evident, intensified. Pressures of a rather different sort grew after 1979 and especially after the 1983 general election as central government sought to privatize activities such as catering and cleaning, especially from the National Health Service, leading to employment reductions and deterioration in working conditions and the quality of services provided.

The third area in which central government public expenditure cuts had an important effect in the North-east was regional policy expenditures. This reduction, moreover, occurred alongside growing competition from other countries for new manufacturing investment and the removal of capital export controls by the Thatcher government in 1979. By the second half of the 1970s the inflow of established companies and branch plants had virtually dried up (Tyne and Wear County Council, 1983). Nevertheless existing big private sector companies already established in the region continued to use regional policy grants to restructure and shed jobs to retain competitiveness. By the late 1970s the pace of job shedding accelerated as companies switched from expansion to capacity contraction as the pressures generated by government policies and international competition intensified (Beynon *et al.*, 1985). This further compounded the effects of employment losses associated with reduced production or total closure of both plants in 'old' industries and the 'new' branch plants earlier located in the region as the effects of the accelerating hypermobility of capital (Damette, 1980) were experienced in the labour market in the North-east.

And as the competition intensified between areas for the new investment which was being made, the battle in the market-place intensified in what was now an international market-place for production sites (Hudson, 1986). It seems that the North-east can compete effectively on this global terrain, most visibly in its much publicized success in attracting inward investment from Japan, symbolized by Nissan's car assembly plant at Washington New Town. This success is deeply ambivalent, however, in terms of its implications for and impacts upon labour markets both in the North-east and elsewhere. Companies establishing new production facilities in the North-east can be extremely selective in their recruitment practices because of the massive pool of unemployed. Nissan received over 10,000 applications for the first 240 jobs advertised at its Washington factory (see *Newcastle Journal*, 8 September 1986). Moreover, they can specify the precise sorts of one-union or no union arrangements and terms and conditions of employment that they want. Trades unions willingly embraced this as the route to some new jobs, not least because attempts to save existing jobs by contesting major closures failed. So leading trades union figures became involved in actively promoting the attractions of the adaptability and non-militancy of the region's labour force, to help attract inward investment. Even so, new companies often systematically avoid recruiting from the registered unemployed, preferring people already working for other companies or coming onto the labour market for the first time, without experience of working in a factory.

Nevertheless, the very fact that companies like Nissan can recruit so selectively emphasizes the extent to which they barely scratch the surface of the unemployment problem. In recognition of this, from the mid-1970s there has been a boom in central government temporary job schemes and training and re-training programmes, especially for school leavers and the long-term unemployed (Manpower Services Commission Northern Region, 1983). As a result, in places such as Consett the main feature of the labour market is that Manpower Services Commission has become the chief employer (Robinson and Sadler, 1984; Boulding, Hudson and Sadler, 1988), as state management of the unemployed has taken on qualitatively new dimensions. From the 1970s there were also growing pressures on local authorities to expand or develop their own 'autonomist' economic development and employment programmes, principally to encourage the growth of 'local' small firms (Northern Region Strategy Team, 1977; Hudson, 1988, chapters 11 and 12).

Such local initiatives have produced few jobs (Storey, 1982), however, and have had at best a marginal impact on the unemployment total. How then have people in the North-east responded to becoming unemployed in a mass unemployment region? Although net annual out-migration has averaged 9,000 since 1977 (Department of Trade and Industry, 1986), inter-regional migration within the UK is no longer a feasible option for most, principally because of house price differentials between the North-east and the South-east (see Hudson and Williams, 1986). For some, with skills in demand in other parts of the world, becoming a temporary international migrant labourer offers one source of wage income (Beynon, Hudson and Sadler, 1985). For others, casual employment within the North-east has been a possibility as established employers restructure their workforces to give a smaller 'core' permanent workforce with 'peripheral' workers hired and fired on short-term contracts as and when required. This option, however, provides irregular and unpredictable wage income. Often the push towards sub-contracting and use of casual labour has come from nationalized industries. Initially, trades unions resisted the more extreme of these proposals but pressures to accept any sort of paid work in a mass unemployment economy produced a reluctant acceptance of them in public and private sectors alike (Hudson and Sadler, 1985; Beynon *et al.*, 1985). For yet others, participation in a 'black economy', sometimes involving home-working, offers possibilities but these are limited in a depressed economy such as that of the North-east.

The collapse of the region's economy had profound effects on housing provision, especially in the context of severe cut-backs in public sector construction because of expenditure cuts. Many local authorities also faced massive bills to repair the deteriorating 1960s system-built estates precisely at the time when central government further tightened its grip over their spending, and when the accelerating closure of collieries and factories was reducing their income from rates. For example, Easington and Sedgefield district councils took over the housing stocks of Peterlee and Aycliffe Development Corporations, respectively, in 1978 – along with a repair bill estimated to be anywhere between £30 million and £60 million. In some cases, local authorities chose to demolish system-built estates because of their unpopularity with tenants and the severity and cost of repair problems. For example, Sedgefield district council pulled down an entire section of the Bessemer Park Estate at Spennymoor in 1983 (*Newcastle Journal*, 29 August 1983), while in 1986 North Tyneside district council decided to flatten the notorious Killingworth Towers, not least because 7,000 families preferred to remain on its housing list rather than accept a tenancy there, correctly identifying it as a 'sink estate' stigmatized by the council's allocation policies. While removing such estates from the public sector housing stock is in many ways desirable, the inability of local authorities to finance new starts results in a reduction of that stock and further tilts the balance in housing provision towards the market.

In other respects, too, local authority policies towards housing provision have of necessity been altered. County council settlement policies have been redefined and much less rigidly applied. This is partly in response to the greater weight attached to private sector construction and owner occupation but also because it is clear that assembling pools of labour for mobile branch plants via local authority planning and construction policies is now an irrelevance. With mass unemployment endemic throughout the North-east, companies experience little difficulty in securing labour-power, wherever they locate within the region. Any pretence that the state can co-ordinate changes in labour and housing markets within the North-east has been dropped, blown away by a return to generalized mass unemployment and a confirmation of the region's marginal position relative to the main currents of accumulation in the global economy.

For many within the North-east buying a house via mortgage credit, redundancy has meant a severe crisis and dislocation in lifestyle, in extreme cases leading to repossession of houses by building societies but more generally leading to a gradual lowering of living standards and, often, physical and mental ill-health. Households in which wage earners lose their jobs have to re-define their aspirations and, often, their position in the housing market, for example, by moving to a smaller house nearer to inner urban areas or from owner occupation back into local authority renting (Beynon *et al.*, 1987). In contrast, in many localities redundancy has had a very different effect on the individual's location in the housing market. Redundancy money has opened the door to owner occupation as sitting tenants have used it to buy their houses from local authorities compelled to sell off their housing stock at substantial discounts by central government, or from their former employer, especially the NCB in (ex) pit-villages. As the push to privatized housing provision has accelerated in these old mining villages and manufacturing towns, it has often occurred amidst numerous boarded up and derelict houses as the precipitate decline in labour demand has rendered parts of the housing stock as redundant as the workers who once lived in it. Even so, buying their house cheaply has provided a degree of security for those nearing retirement age, in terms of their perception of the situation, for they have no wish to leave communities in which they have lived for many years. For others, however, who are younger, inability to sell as the local housing market has slumped in the wake of mass redundancies, especially in what were formerly one-industry communities, has ruled out the migration option, even if they want to follow it and could find some way of coping with house price differentials between the North-east and the South-east. For those who have never left or have returned to the public sector, the reduction in local authorities housing stock and scope for transfers has had the same effect.

More and more, the remaining local authority stock in the North-east has deteriorated in quality because of the impacts of public expenditure policies on repairs. For example, by 1983 almost a third of Newcastle's local authority houses needed a 'complete overhaul' (*Newcastle Journal*, 31 August 1983). Local authority housing has become – *de facto* – the location for spatially demarcated concentrations of the unemployed, a burgeoning group of 'clients of the welfare state' largely supported by state transfer payments but no longer confined to a few administratively selected 'sink estates' as in the 1960s as unemployment has become the dominant characteristic of the region's labour market. It is not only the unemployed who are trapped in these public sector estates, however, for those working in and on the fringes the black economy are unlikely to meet the criteria laid down for access to mortgage credit and owner occupation.

In sharp contrast, there is a section of the middle and affluent working classes who are in regular waged work, especially in households with two (or more) incomes for whom housing conditions and position in the housing market are markedly different. This is true, for example, of those members of the 'new' middle class employed in the state sector on salary scales tied to national wage norms in a region where, compared to the South-east, housing costs are markedly lower. Likewise, those members of the skilled manual working class able to sell their labour-power on the international market in activities such as construction, chemicals and oil at prices far above those they ever dreamed were possible in the North-east, have often made quite dramatic jumps in position in the housing market. While very different from the speculative boom conditions of the housing market in much of the South-east, there is nonetheless a steadily buoyant market for owner occupied four or more bedroomed detached houses in what are regarded as better, often semi-rural, residential environments. In these localities, two car households, equipped with the latest in dishwashers, freezers and video recorders and a lifestyle facilitated by credit provided are, if not quite the norm, certainly not uncommon. The contrast with the poverty and material deprivation of the rotting council estates of the inner areas of Teesside, Tyneside and Wearside could not be sharper, a poignant reminder that locations in and trajectories through housing and labour markets tend to be mutually reinforcing and indicative of increasing social polarization within the North-east as well as between it and the South-east.

Summary and some conclusions

As its baldest, the contrast between state policies and changes in housing provision and labour markets may be summarized as follows. In the period 1942–58 there was a powerful consensus around the need to stabilize the labour market for men while encouraging some sectorally and locationally selective increases in female wage labour. Local authority housing and planning policies were formulated so as to try simultaneously to support both of these forms of labour market stasis and change via locationally concentrated provision of new public sector housing. By and large, patterns of labour market stability and change and changed patterns of housing provision were engineered as intended. Even so, there were clear signs of definite limits to this process. In particular, growing participation of (married) women in waged labour provided a mechanism for increasing household expenditure on better quality public sector housing without substantial increases in male wages. Yet there were limits to married women working outside the home for a wage as long as the pattern of male shift-working in industries such as coal and steel continued and in a cultural context which prioritized the role of the man as the household wage earner.

After 1958 these limits began to be eroded rapidly. The priority altered from preserving to reducing male labour forces in 'traditional' industries. The emphasis in state policies was directed towards attracting alternative male employment and new forms of regional planning and policy were introduced to try and do so. A broad consensus emerged around the need to modernize both environment and people, if modern industries and new employment were to be attracted to the North-east. It was also accepted that because of public expenditure constraints, modernization policies would need to be focused upon selected growth points within the region, into which the new houses and factories would be channelled in an inter-related and locationally concentrated transformation of labour and housing market conditions. In contrast to the previous period, greater emphasis was now placed upon encouraging owner occupation, a more modern form of housing provision in these places. Many of these planned changes in environments, housing and living conditions proceeded as intended, unlike the associated labour market changes. The volume of new jobs fell far below that of old ones that were lost and there were serious disjunctions in the gender and skill composition of jobs gained and lost. About half of the new manufacturing jobs and most of those in the service sector, private and public (many on a part-time basis), were taken by women. As the female wage labour force grew, male unemployment rose. These divergences between intended and actual labour market changes had implications for changes in the pattern of housing provision. What was beginning to emerge was a growing polarization between households with two (or more) wage incomes, living 'modern' lifestyles in owner occupied housing and relative affluence, and those households with no wage earners, reliant upon state transfer payments for a minimal level of subsistence and confined to local authority 'sink estates' by housing allocation procedures, where they could be more easily surveilled by the police and social security officials.

As unemployment became more and more generalized after 1975, this polarization intensified. Local authorities have accepted the futility of trying to co-ordinate labour and housing market changes via settlement policies in the face of central government's commitment to market forces. As job losses continued in coalmining and in 'new' as well as 'old' manufacturing industries, women as well as men lost their wage incomes. Furthermore, as the effects of downward multipliers and public expenditure cuts worked through to the service sector, the rate of job growth slowed to zero, then became negative, with part-time jobs increasingly replacing full-time ones for women. For those remaining in formal waged work or obtaining one of the few new jobs on offer in the market, working conditions and practices were often sharply redefined while around these reduced core workforces, employment on a sub-contracted or casual basis, often in or on the fringes of the black economy, expanded. Some who had lost their jobs used their redundancy money to finance a switch to owner occupation, typically in their now deindustrialized former pit village or steel town, but the majority without formal waged incomes remained trapped in or returned to a shrinking stock of deteriorating public sector housing. The role of the state increasingly became management of the unemployed, via temporary job and training schemes as well as housing allocation policies. In contrast, those remaining in waged work, especially the middle class in two income households where wages were tied to national norms, had access to a very different quality of owner occupied housing and lifestyle.

Finally, it should be noted that the residual of the landed aristocracy and major capitalists within the North-east (Benwell Community Development Project, 1978a) retained a lifestyle that was insulated from the vagaries of the labour market, typically living in rural or semi-rural locations away from the coalfield and the major conurbations.

What conclusions are to be drawn from this sorry tale of labour and housing market change? Perhaps the most obvious is that changed positions in the two markets are related in various ways. The state was unable to manage these changes as it intended even when its policies were informed by progressive social democratic concerns; and now there has been a switch to a minimal concern for progressively managing change via state policies. The rhetoric of the market as the mechanism for allocating houses and jobs has powerfully re-emerged. There is a connection here. For what became clear in the 1960s was that the (implicit) theory of the state as a set of neutral instruments to be captured and turned to social democratic purposes was revealed as an inadequate basis for practice. Above all, the growing internationalization of the economy cut the ground from national policies of economic management and planned labour market changes in regions such as the North-east. This, in turn, undermined planned changes in housing provision so that the actual patterns of housing and labour market change bore only a partial resemblance to those intended. To reduce this divergence, however, would have required mounting a much stronger attack against capitalist property relations and the political basis to reformulate state policies along these lines simply did not exist. It is in these circumstances that the emergence of a more market-oriented politics from the mid-1970s must be understood, a switch that further polarized inter-related positions in housing and labour markets within the North-east as well as between it and the South-east.

References

Austrin, T. and Beynon, H. (1979), *Global Outpost: The Working Class Experience of Big Business in the North East of England, 1964–79* (Department of Sociology, University of Durham).

Benwell Community Development Project (1978a), *The Making of a Ruling Class*, Final Report No. 6, Benwell CDP 85/7 (Adelaide Terrace, Newcastle upon Tyne).

Benwell Community Development Project (1978b), *Slums on the Drawing Board*, Final Report Series No. 7, Benwell CDP 85/7 (Adelaide Terrace, Newcastle upon Tyne).

Beynon, H., Hudson R. and Sadler, D. (1985), *Mis-managing Horden*, Work and Employment Research Unit, Durham University.

Beynon, H., Hudson, R. and Sadler, D. (1986a), 'Nationalized industries and the destruction of communities: some evidence from North-east England', *Capital and Class*, 29, pp. 27–57.

Beynon, H., Hudson, R. and Sadler, D. (1986b) 'The growth and internationalization of Teesside's chemicals industry', *Middlesbrough Locality Study, Working Paper No. 3*, University of Durham.

Beynon, H., Hudson, R., Lewis, J., Sadler, D. and Townsend, A. R. (1985), 'Middlesbrough: contradictions in economic and cultural change', *Middlesbrough Locality Study, Working Paper No. 1*, University of Durham.

Beynon, H., Hudson, R., Lewis, J., Sadler, D. and Townsend, A. R. (1987), 'A changing locality in an international economy: the contradictions of change in Middlesbrough', in P. Cooke (ed.), *Global Change, Local Adjustment* (London, Hutchinson).

Board of Trade (1963), *The North East: a Programme for Regional Development and Growth*, Cmnd 2206 (London, HMSO).

Boulding, P., Hudson, R. and Sadler, D. (1988), 'Consett and Corby: what kind of new future?', *Urban Affairs Quarterly*.

Byrne, D. and Parson, D. (1983), 'The State and the reserve army: the management of class relations in space', in J. Anderson, S. Duncan and R. Hudson (eds), *Redundant Spaces in Cities and Regions*? (London, Academic Press), pp. 127–54.

Cullingworth, J. B. and Karn, V. A. (1968), *The Ownership and Management of Housing in New Towns* (London, HMSO).

Damette, F. (1980), 'The regional framework of monopoly exploitation', in J. Carney, R. Hudson and J. Lewis (eds), *Regions in Crisis* (London, Croom Helm), pp. 76–92.

Department of Trade and Industry (1986), *UK Regional Development Programme 1986–90 Section 7A: England – North East* (London, DTI).

Horner, A. (1960), *Incorrigible Rebel* (London, MacGibbon and Kee).

Hudson, R. (1980a) 'Women and work: a study of Washington New

Town', *Occasional Publication No. 16*, Department of Geography, University of Durham.

Hudson, R. (1980b), 'Regional development policies and female employment', *Area*, 12, 3, pp. 229–34.

Hudson, R. *et al.* (1976), *New Towns in North East England*, Final Report HR 1734 to the Social Science Research Council (London).

Hudson, R. (1982a), 'Accumulation, spatial policies and the production of regional labour reserves: a study of Washington New Town', *Environment and Planning A*, 14, pp. 665–80.

Hudson, R. (1982b), 'Capital accumulation and chemicals production in Europe in the post-war period', *Environment and Planning A*, 15, pp. 105–22.

Hudson, R. (1985) 'The paradoxes of State intervention: the impact of nationalized industry policies and regional policies in the Northern Region in the post-war period', in R. Chapman (ed.), *Public Policy Studies: North East England* (Edinburgh, Edinburgh University Press).

Hudson, R. (1986), 'Producing an industrial wasteland: capital, labour and the State in North East England', in R. L. Martin and B. Rowthorne (eds), *Deindustrialization and the British Space Economy* (London, Macmillan).

Hudson, R. (1988), *State policies, party politics and regional change* (London, Pion).

Hudson, R. and Sadler, D. (1983), 'The closure of Consett steelworks: anatomy of a disaster', *Northern Economic Review*, 6, pp. 2–17.

Hudson, R. and Sadler, D. (1984), *British Steel Builds the New Teesside?*, Cleveland County Council, Middlesbrough.

Hudson, R. and Sadler, D. (1985), 'The development of Middlesbrough's iron and steel industry, 1841–1945', *Middlesbrough Locality Study, Working Paper no. 2.*

Hudson, R. and Williams A. (1986), *The United Kingdom* (London, Harper and Row).

Hudson, R., Peck, F. and Sadler, D. (1985), *Undermining Easington?*, Easington District Council, Easington, County Durham.

Mandel, E. (1978), *The Second Slump* (London, New Left Books).

Manpower Services Commission Northern Region (1983), *REMIT*, Newcastle upon Tyne.

Newcastle Journal, 23 June 1983; 29 and 31 August 1983; 26 November 1985; 8 September 1986.

North Tyneside Community Development Project (1978), *North Shields: Women's Work* (London, Home Office).

North East Trades Union Studies Information Unit (1977), *Multinationals in Tyne and Wear*, vols. 1 and 2, Newcastle.

Northern Regional Strategy Team (1975), 'Change and efficiency in manufacturing industry in the Northern Region', *Technical Report No. 3*, Newcastle upon Tyne.

Northern Region Strategy Team (1976a), 'The chemical industry in the Northern Region', *Working Paper No. 1*, Newcastle upon Tyne.

Northern Region Strategy Team (1976b), 'Linkages in the Northern Region', *Working Report No. 6*, Newcastle upon Tyne.
Northern Region Strategy Team (1977), *Strategic Plan for the Northern Region*, vols. 1–5 (London, HMSO).

Pepler, G. and MacFarlane, P. W. (1949), *The North East Area Development Plan*. Interim Report presented to the Minister of Town and Country Planning.

Robinson, J. F. F. (1978), *Peterlee: a study of New Town development*, unpublished PhD thesis, University of Durham.
Robinson, J. F. F. and Sadler, D. (1984), 'Consett after the closure', *Occasional Publication (New Series) No. 19*, Department of Geography, University of Durham.
Rowntree Research Unit (n.d.), 'The origins and development of Cramlington New Town – a Chronology', unpublished manuscript, University of Durham.

Short, J. R. (1982), *Housing in Britain* (London, Methuen).
Storey, D. (1982), *Entrepreneurship and the New Firm* (London, Croom Helm).

Taylor, P. J. (1979), 'Difficult to let', 'difficult to live in', and sometimes difficult to get out of': an essay on provision of council housing, with special reference to Killingworth', *Environment and Planning A*, 11, pp. 1305–20.
Tyne and Wear County Council (1983), *Economic Development Bulletin, No. 17*, Newcastle upon Tyne.

9 *Housing the workers in Mrs Thatcher's hi-tech Utopia: housing markets and labour markets in Berkshire*[1]

JAMES BARLOW & MIKE SAVAGE

Localized employment growth poses specific problems for any system of housing provision which is heavily reliant on speculative private sector housing provision. New workers need somewhere to live, hence there will potentially be a demand for new housing, with possible effects on house prices and rents. In short, the potential contradictions in such a system of housing provision are likely to be especially exposed in areas of employment growth.

In this chapter we will demonstrate this process by a detailed study of changes in the labour and housing markets in Berkshire, England. This area lies to the west of London and is part of the 'M4 Corridor' (see Figure 9.1). It has seen considerable recent economic growth, partly through decentralization of economic activity from London and partly through the development of indigenous private firms, especially in the service and 'hi-tech' sector. It has also been one of the major areas for speculative housebuilding for owner occupation in the past decade. In this chapter we will argue that there is, however, no smooth relationship whereby demand associated with economic growth leads to more and better housing.

We begin by probing behind the image of economic growth in Berkshire. Although there has been net employment growth, some groups have benefited more than others, and it is misleading to paint a picture of 'prosperity for all'. We then move on to discuss how forms of housing provision are changing. Here we will see that the dominant form of housing development in Berkshire (speculative owner occupation) tends to accentuate divisions already opening up in the labour market, and hence helps to increase social polarization within the county. We will conclude by showing how these are fuelling political and social tensions more generally.

Restructuring the economy

Superficially Berkshire is a prosperous county, one with an affluent expanding economy well able to afford owner occupied housing. As *The Economist* (30 January 1982) put it, 'Berkshire was the last bit of Britain into the recession. Signs are it will be the first out . . . Britain needs more Berkshires.' This economic expansion has, however, not been without certain traumas. Alongside the rapid growth in services there has been a considerable decline of industrial employment in the area. While these two trends may seem to cancel each other out, there is relatively little interplay between them: workers displaced in manufacturing rarely move into the expanding sectors (see Savage, 1987a).

Berkshire has a certain reputation as a genteel, rural county, uncontaminated by industry, but such a picture fails to convey its central position in the development of the new 'mass production' industries of the interwar years (especially in motor cars, electrical engineering and food processing). This was especially true in the case of Slough (actually in Buckinghamshire until 1974). Here, the Slough Trading Estate Ltd perfected a new scheme to develop industry (modelled on the pioneering Trafford Park Estate in Manchester) whereby the company would provide land, factory premises, energy and even housing for employees, and would rent out these facilities to other manufacturing companies. This system suited perfectly the requirements of a number of multinational firms moving into Britain, who knew little about the precise nature of specific sites, and wished to move into ready-made factory units.

These firms specialized in semi-skilled, labour intensive, mass production processes. A large industrial working class was formed through an influx of workers from the 'distressed' areas (especially South Wales), the tapping of the remnants of the old agricultural labour force and through the widespread use of female and youth labour (Savage, 1988). After World War II the New Town of Bracknell became the main growth point, especially for electrical engineering, and in Reading the traditional reliance on 'beer, bulbs, and biscuits' (Courage Breweries, Suttons' Seeds and Huntley and Palmers) was supplemented by newer engineering concerns. In 1961 Berkshire had a slightly higher proportion of its workforce in electrical engineering, food processing, and paper and printing than did England and Wales as a whole.

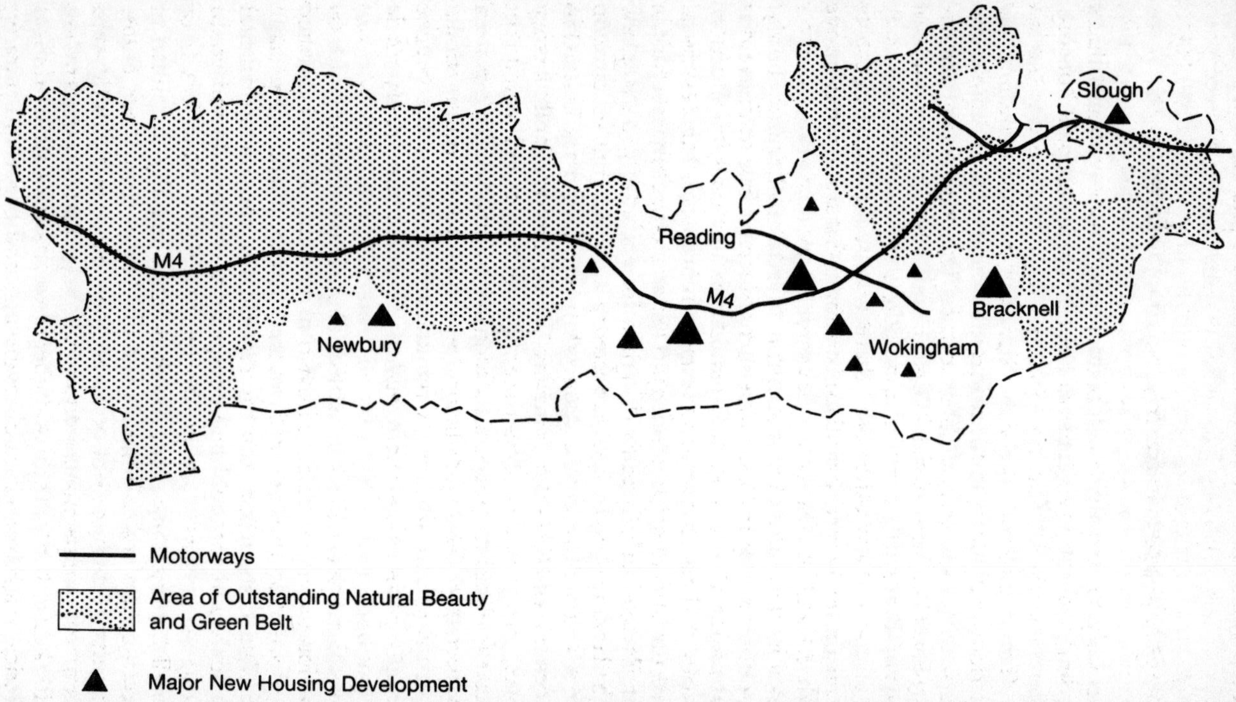

Figure 9.1 The 'M4 Corridor' and new housing in Berkshire

Expansion continued apace into the 1970s. Employment rose in the 1970s by 2 per cent a year in Berkshire, against only 1.3 per cent in the rest of the relatively buoyant South-east region. The recession appeared to pass Berkshire by as employment rose from 310,000 in 1976 to 340,000 in 1984. Yet, it is important to recognize that many of the older, mass production industries have declined, just as they have elsewhere, but that in Berkshire they have been replaced by new sources of employment based on private services and 'hi-tech' industries.

'Hi-tech' growth started in the late 1950s and was based initially on large, often multinational, firms, many diversifying from mainstream electrical engineering. These firms include Hewlett Packard, Racal, Ferranti, Honeywell, Digital and ICL, and were based mainly in Bracknell. Recently, though, there has been a rise in small indigenous hi-tech firms, often set up by previous employees of the larger firms, which spread out to other parts of Berkshire. By the 1980s about 30,000 people, just over 10 per cent of the county's workforce, were employed in the hi-tech sector (Morgan and Sayer 1988; Hall *et al.*, 1987). Although the expansion of electronics firms within the area has been well publicized, the growth of private producer services has perhaps been of more importance. Private services (including some 'hi-tech' activities such as software) are in fact the most dynamic sector of Berkshire's economy: 37,000 new jobs were created in Berkshire between 1971 and 1981, a growth rate of 46 per cent. This increase was more than were employed in the entire 'hi-tech' sector in 1981 (RCB, 1985a; see also Hall *et al.*, 1987).

The crucial point is that in these growth sectors there has been relatively little demand for the local residents who have historically found employment in the older manufacturing sectors. Many of the jobs in hi-tech firms rely on mainly male professional workers, and offer few opportunities for manual workers. Only 12 per cent of jobs in this sector are for unskilled labourers, compared with 29 per cent in other Berkshire firms. No less than 55 per cent are for professional or managerial workers (RCB, 1985b). The large hi-tech firms, especially those in Bracknell, continue to import many of their managerial and professional workers from outside the area.

Most manufacturing firms have survived, but they have had to improve productivity. This has usually been achieved by introducing new technology, such as at Mars in Slough, using subcontractors (there are many 'jobbing shops' on the Slough Trading Estate), and generally cutting their 'core workforce'. Sometimes, as with Mars, these are replaced by workers on short-term contracts. There have been extensive job losses in the older manufacturing sector. Between 1979 and 1983 unemployment in Berkshire rose from 2 per cent to 8 per cent, and 75 per cent of the jobs lost were in manufacturing. By 1984, unemployment in Reading was 11.4 per cent, only a little below the national average. Many of the older firms on the Slough Trading Estate have moved out: High Duty Alloys, which employed several thousand in the postwar period is the most notable example. On a smaller scale, Black and Decker have moved their distribution to Northampton, and Berlei are moving warehousing to Wales. Slough Estates has in fact chosen to redevelop much of the Trading Estate to attract new hi-tech and service sector firms. Other firms who owned their freehold have chosen to sell highly valued inner city sites (particularly in Reading) for office development.

There is, however, little evidence that old manual manufacturing workers have found work in the expanding sectors (Savage, 1987a). Green (1985) shows that although employment in services grew twice as fast as the national average between 1971 and 1981, much of the increase was in the managerial ranks. The number of skilled, semi-skilled and unskilled manual workers in services actually fell in Berkshire between 1971 and 1981, in most cases more than the national average, while the numbers of employers and managers in services was rising at over twice the national rate (Green, 1985, Table XI). Furthermore, much of the employment increase went either to men or part-time women employees, and only in the declining public distribution sector were the proportions of full-time women workers increasing.

After the Second World War many women had remained active in the local labour market, encouraged by manufacturing employers suffering from severe labour shortages. Women's participation rates in Berkshire have remained above the national average, with rapidly growing employment in the private service sector. However, these jobs also tend to be casual and not conducive to long-term career development, especially compared with the public sector (Crompton and Jones, 1984; Savage, 1987b).

To summarize, although in aggregate terms there has been job growth, many local people – especially male manual workers – have suffered redundancy in Berkshire (as elsewhere in Britain) and have rarely been able to find equivalent secure and remunerative employment. Their capacity to engage in the owner-occupied housing market has also been progressively undermined at the same time that many of the expanding sectors have drawn an affluent, professionalized workforce into the area, which in turn has set up demands for certain forms of housing. Some of these workers are part of large corporate internal labour markets – the types of workers who can draw on company assistance for mortgage payment (see Chapters 2 and 3; Barlow and Savage 1986, 1987). We should be careful not to overemphasize these groups since their mobility into the South-east from other regions of the UK has been declining (see Savage, 1987c; Savage, Dickens and Fielding, 1988). More importantly, many of these workers have moved out of London as part of the trend towards decentralization, and in search of larger 'family' housing in a 'better environment' (Barlow, 1988a).

Restructuring housing provision

Just as Berkshire has been put forward as a model of Britain's industrial future, so its distinctive housing developments have marked a great departure from past practice. On the surface it may appear that the whole county's population has benefited from the remarkable expansion of the housing stock in recent years. Between 1976 and 1984 some 38,000 houses were built, a 17 per cent increase in Berkshire's housing stock (RCB, 1985c). In March 1985 there were outstanding commitments for a further 23,000 houses, 79 per cent of which were in central Berkshire, many around the district of Wokingham (RCB, 1985d). Virtually all of the houses have been built speculatively by private builders for owner occupation, often on very large sites such as Lower Earley, where permission has been given for 6,000 houses (RCB, 1985d). These developments have changed the tenure mix of the county. In 1971 56.7 per cent of Berkshire's residents were in owner-occupied property, but by 1981 this figure had risen to 66.7 per cent.

Developers, often the large volume builders, have been very successful in pushing for land release in large blocks, and have pressurized central government to over-ride the more conservationist-minded local authorities. In the early 1980s, Michael Heseltine, then Secretary of State for the Environment, modified the Central Berkshire Structure Plan to release land for an extra 8,000 dwellings (RCB, 1983; Fleming, 1984; Witt and Fleming, 1984). The plans for a number of 'Heseltowns' were incorporated into the Stage Two Structure Plan proposals. In January 1988 Nicholas Ridley, the then environment minister, ordered the county council to release land for an extra 6,970 houses, over and above the Structure Plan target (the *Independent*, 30 January 1988; the *Guardian*, 24 February 1988).

In order to understand the changes in the housing structure of Berkshire we need to consider some of the trends evident in the British housing provision system during the 1970s and 1980s. There are several important features of this period (Barlow, 1986, 1987).

First, although the total amount of central and local government expenditure on housing provision has actually grown in constant prices since 1978 and 1979, there has been a substantial shift in emphasis away from direct investment in new public sector housing towards support for housing consumption, via mortgage interest tax relief and housing benefit. Parallel to the decline in new public sector construction there has been the sale of over 600,000 council houses and restrictions on local authority capital expenditure (Forrest and Murie, 1986). Second, there has been a growth in indirect support for private sector development in the form of Urban Development Grants and Derelict Land Grants, although nationally the impact of this type of support has as yet been limited. Third, the private sector developers have reoriented their operations towards the more affluent regions. The South-east's share of total new private housing starts has risen steadily from about 28 per cent in 1974 to nearly 39 per cent in 1986. The developers have also sought new types of market, particularly 'trade-up' customers and the retired.

Massive housebuilding has not eased the problems of many Berkshire residents, though. For the housebuilders the incoming professional workers have provided an ideal 'trade-up' market. Many of the households who have moved into the county have migrated from London, and therefore can afford relatively expensive housing. The average weekly household income of local borrowers from the Nationwide Building Society in 1984 was £307 (NBS, 1985), compared to an average in the South-east (excluding London) of £242. Not surprisingly, the size of mortgage given to first-time buyers is generally bigger than that given to previous owner occupiers, the reverse of the national pattern (RCB, 1985b). It has also been estimated that two-thirds of purchasers of new dwellings in 1983 and 1984 were professional or managerial workers (RCB, n.d.). In the light of these facts, it is not altogether surprising that house prices have increased faster than in the rest of the South-east. In 1984 the average price of all houses in Berkshire was almost £49,000, compared with £40,000 in the rest of the South-east and £31,000 nationally (RCB, 1985b).

Yet as these changes have taken place, the security offered by the public sector has been rapidly eroded by the dramatic cutback in local authority housing provision. Bracknell has been a notable case. This town had a large amount of high quality housing built by the New Town Corporation, which proved highly popular among purchasers. Almost one-third of the public housing stock, 3,400 houses, were sold off between 1981 and 1984. As the stock dwindled, the local council, under Tory control, had restricted single people's eligibility for council housing, and in 1981 applicants had to be aged over twenty-five before being placed on the waiting list. There has also been a high rate of council house sales in Slough, although this town has also had an extensive programme of new building. In general, though, council housing is not being replaced: Berkshire currently has the lowest proportion of public sector completions in the South-east. This has led a number of firms to complain about the lack of council housing for their workforces (RCB, 1985a).

It is also the case that areas of urban deprivation have been untouched by the housing developments springing up only a few miles away. As the *Guardian* (2 October 1985) reported, 18 per cent of the households in the Battle Ward in Reading lack their own bath or inside toilet. The County Council's 'league table' of economically deprived wards shows a concentration in Reading, Slough and Bracknell.

Many sections of the local population, especially the single and young, those on low incomes and one-parent families, have been effectively 'shut-out' of the housing market. Levels of homelessness have risen considerably, and there has been an expansion in 'bed and breakfast' establishments (private rented accommodation funded by social security payments). In 1981 85 per cent of professional, administrative and managerial workers (Socio-economic Groups 1–4) were owner occupiers, compared to only 51 per cent of the manual working class (SEGs 9–11) (Census, 1981). In 1984 in Wokingham, the district with the highest rates of housebuilding, only 22.8 per cent of Nationwide borrowers were first-time buyers compared to the national average of 42.1 per cent (NBS, 1985).

Political and social tensions

It appears that the 'privatized', market-oriented housing system has accentuated growing social and economic tensions in Berkshire. The system is failing to fulfill the housing needs of large segments of the population, and is also creating other political problems, notably around the rising cost of servicing the growth and in the politics of growth.

First, the local authorities have been hit by the need to provide an infrastructure for these new housing developments, while facing restrictions on local government expenditure. It has been estimated (RCB, 1985c) that the capital cost of providing services for development already committed will be some £29 million per annum up to 1996, with the government permitted maximum running at £25 million. In addition, Berkshire will need to find an extra

£16 million in running costs and £31 million in loan charges every year. The county council's outstanding loan debt was predicted to rise to £191 million by April 1987 (RCB, 1986). Under the present expenditure constraints imposed by central government on local government it is unlikely that any extra money will be forthcoming. As the Structure Plan Review notes, 'there will be even less scope for dealing with current problems . . . unless the Government makes more financial resources available and/or developers meet a larger share of the costs of providing infrastructure needed as a result of development' (RCB, 1985c, para. 2.3.10). Government expenditure targets, it adds, 'have tended to bear harshly on counties such as Berkshire, where population is increasing' (RCB 1985c, para. 6.2.3). Essential roads and services for the Lower Earley development were partly funded by the housebuilders, although the money was not available to the local authority until the housing was sold. More recently, there has been a move towards insisting that developers provide the physical infrastructure rather than lump-sum financial payments. While the general policy of the Thames Water Authority is to cater for growth and charge builders for work done, it too has found that cuts in its capital programme have led to delays in some projected developments.

It is hard to say whether future growth is threatened by a lack of cash for infrastructure investment. It is possible that housebuilders will pass on the extra costs imposed on them to house purchasers, or that they will attempt to depress the price offered to landowners. What is clear, however, is that the existing infrastructure, particularly transport facilities, has been seriously deteriorating in recent years, especially as the rate of car ownership has now risen above one car per household. Yet the county spends only half as much money per head on public transport and roads as most of the former metropolitan counties (with which Berkshire is comparable because of its high population density). Water and health services are also suffering from similar constraints. Green (1985) has shown that between 1978 and 1981 employment in public services in Berkshire fell by 2.45 per cent against a figure of 0.89 per cent nationally.

Second, the influx of professional and managerial workers, looking for their 'place in the country', and the attempts by speculative housebuilders to cater for this market have tended to compound social tensions in the area. The conservationist lobby has become extremely vociferous with its most notable success being the halting of the county council's plan to allow 4,000 extra houses in north Bracknell. Many of the most active conservationists are people who moved to a 'rural' area under the impression that their development was the last which would be permitted, only to find housebuilders (having charged higher house prices for the 'rural' location) pressing for yet more land release. The growth of professional and managerial workers chasing housing of the appropriate status, is one of the major causes of local political cleavage, as they act to try to prevent further development (see Barlow and Savage, 1986 for further details; Saxenian, 1985 for some US parallels). The county council takes a firm anti-growth stance and the Structure Plan Consultation Document argues that too much land has already been released. The approved strategy is now to 'preserve the rural character of the County as a whole and to direct such development as is necessary to those areas which can best accommodate it and where services can most easily be provided' (RCB, 1985c, para. 4.2.1). This did not, however, seem to have caught the ear of Nicholas Ridley (see above).

All three political parties now agree that some form of control over development is needed, and although there is disagreement about how and where the reduction should occur, the consensus is that growth should be limited to central Berkshire. This does not square with the demands of the housebuilders, though. Over the last few years the builders, the planners in the counties of South-east England and the Department of the Environment have been arguing over future housebuilding needs and the supply of land. Berkshire has been a prime battleground. The housebuilders have been actively lobbying the DOE for revised estimates of future population change. Yet Berkshire county council planners argue that DOE projections have been based on faulty assumptions about rates of household formation, since these have failed to take into account the reductions in the number of households formed caused by the recession. Furthermore, almost half of new household formation will be of single households, yet this is not the type of demand which the housebuilders are planning to cater for. While the planners in Berkshire argue that the rate of population growth will be primarily determined by their planning policies, such views do not square with notions of letting the 'free-market' determine the speed and location of growth. Apart from questioning the population forecasts the housebuilders have countered that such countywide estimates of land supply fail to identify local imbalances in supply and demand.

Berkshire is a key site for conflicts over development, partly because of the rise of its affluent middle class. This has led to rifts within the Conservative Party, with the conservationist issue dividing rather than uniting the Party. Tory MPs, especially Andrew Mackay, MP for Berkshire East, who rely on support from owner occupiers, have been most active over this issue. Mackay had the support of eighty-five Tory MPs in asking a previous environment minister to think again about giving land to builders in Berkshire (the *Sunday Times*, 10 March 1985). Other traditional Tory groups have supported development, though. These include the farming and landowning lobby and the employers. The *Sunday Times* (2 January 1983) reported that 'the farmers and the landowners can hardly wait [for development]. Some fifty-two have already written to the County Council to offer land.' These groups clearly stand to make money from selling land to housebuilders, although there are divisions emerging within the farming sector over land sales (Barlow, 1988b).

There is, then, a complex relationship between the labour and housing market. The growth of professional and managerial workers in the county has led to rising demand for housing. This demand has been strengthened by the fact that many of these workers are in-migrants and few of these workers are recruited from Berkshire's declining industrial sectors. In the present context the housebuilders have clearly found the 'trade up', middle-class market particularly profitable. However, it is these very residents who are often opposed to further housebuilding. As a result it is harder to develop the large sites which might cater for a wider social group than the small-scale infilling which can still be forced through the planning system. Concentrating new housebuilding on smaller sites may, some builders argue, reinforce house price increases as they are unable to cross-subsidize by building a wide range of house types.

County planners agree that the conservationist movement has not come to terms with a basic problem. While they are opposed to *housing* development many councillors are not opposed to further *economic* development. Yet such development would be likely to recruit from professional labour markets, and would therefore result in a further influx of people needing somewhere to live, fuelling house price inflation.

It is evident that the high house prices have also been hitting local employers. Almost a quarter of large firms in a recent survey (RCB, 1985a) indicated that they were experiencing difficulties in housing their workforce, with the cost of housing as the main barrier. The House-Builders' Federation has argued (perhaps not surprisingly) in its submission to the Central Berkshire Structure Plan inquiry that housing shortages have hindered recruitment of skilled, technical and administrative staff and that the cost of housing relative to other parts of Britain is the major problem (Parsons, 1986). Indeed, a number of firms have expressed concern over rising housing costs: National Semiconductors, for example, stated that the company 'would have liked to be nearer Heathrow Airport . . . but Swindon won in the end because of the area's house prices' (Pauley, 1986). The East Berkshire and South Buckinghamshire Chamber of Commerce and Industry (the fifth largest in the country), representing the interests of local manufacturers, supports further housebuilding and argues that there is a chronic shortage of rented accommodation. One official went out of his way to condemn the government regulation whereby councils can only use 20 per cent of the receipts of council house sales to build more council accommodation. This situation is deteriorating as labour shortages in particular sectors become more apparent. The Chamber stated in September 1985 that the labour shortages for skilled manual workers were the worst ever, with 80 per cent of the firms who were looking for such workers experiencing difficulty in recruitment.

Of course, much of the lobbying over the effects of house prices on the labour market may well stem from the volume builders desire for more land. While it is true that house prices are extremely high relative to the national average the effect of high house prices on firms' locational decisions should not, perhaps, be overstated. The variation in housing costs between London and the South-east and the rest of the country in 1986 was no greater than it was during the early 1970s boom. At least one survey has shown that the overwhelming majority of firms (71 per cent) moved primarily to expand their activities, with the average distance moved being only 16.4 kilometres (Brown, 1986). In fact, the campaign is clearly about freeing the housebuilding industry and other industries from planning constraints: statements have emerged from the Planning Minister (Lord Elton) that the prosperity of the 'national economy' is threatened by limitations (caused by planning 'red-tape') to the freedom of London's industry to develop as it needs (Barlow, 1986).

Housing markets and labour markets: some conclusions

The effects of changes in the local economy and in housing provision on social polarization are relatively clear. Within Berkshire social cleavages have been exacerbated by recent developments, and this polarization is manifesting itself through growing local divergences: contrast the poorest wards in the centres of Slough and Reading with the wealthy suburbs of Wokingham, Bracknell and Newbury. Politically, this may be reflected in the fact that Reading and Slough Councils remain Labour controlled (May 1987), while the rural councils around are dominated by the Conservative and Alliance Parties.

In many respects these features are similar to trends found in 'depressed' areas across the country. Goodwin, in Chapter 10, shows how in the London Docklands, an area of massive economic restructuring, the activities of the London Docklands Development Corporation are attracting new firms employing non-local people, and new housing geared to professional markets. The same dynamic is evident in Berkshire. On the one hand there is a highly mobile managerial and professional cadre, frequently moving up the internal labour markets of major corporations, who are well able to use owner occupation to their advantage. These groups, which are rapidly expanding in Berkshire and generally in the South-east have been of critical importance in generating housing demand for which the speculative builders have enthusiastically catered. On the other hand, the local working-class population is increasingly locked into poor housing and insecure jobs, despite the area's relatively buoyant economy.

It is, therefore, important that we treat with caution the notion that working-class owner occupiers are benefiting from expansion.[2] Although capital gains can be made from housing, in most cases this is not disposable income. It is of little use for people in a terraced house in Reading to know that their house is worth £40,000 (compared with £20,000 in a northern conurbation), if they cannot realize this money. This problem is especially acute for young people. It is becoming increasingly difficult for them to enter either the private housing market, or to obtain public sector accommodation. Since the main way in which people can realise money from owner-occupied housing is by trading down once their children have left home, the fact that the children of less well-off people have difficulty in obtaining accommodation adversely affects the potential economic gains from their parents' point of view. It is by no means clear then that the interests of working-class owner occupiers should be against public rented housing.

Analysis of the Berkshire case has also illuminated the relation between the housing market and labour market at a more conceptual level. As Hamnett (1984) and Hamnett and Randolph (1986) suggest, it is untenable to maintain that the housing market operates independently of the labour market. The position of households and groups in these markets can be both mutually reinforcing and mutually contradictory. Thus in Berkshire there seems to be a reciprocal relationship between the housing market and labour market. Changes in the labour market towards a professionalized workforce influences the price of and demand for housing in a given area. As a result the high price of owner-occupied housing which is built to meet this demand makes it difficult for employers to recruit lower-paid workers from outside the area and house lower-paid members of Berkshire's existing workforce. This creates further pressures for firms employing a largely manual workforce to move out of the area, so accentuating the trends leading to the dominance of a high-income professional labour market.

Another example of this process may be the considerable increase in the number of married women at work in the 'Outer South-east' between 1968 and 1977. According to Molho and Elias (1982) this was not caused by the attraction of increasing female wages (as in other areas where their participation rates rose). Rather, it may be that high housing costs in Berkshire and other parts of the Outer South-east are 'pulling' more women into the workforce, as the need for two household incomes to pay the mortgage grows (see Savage, 1987b for a further discussion of this).

Two further points about the labour market/housing market link must be made. First, in spatial terms there is a 'mismatch' between the Berkshire 'labour' market and the Berkshire 'housing' market. On the one hand different places in the county fulfil different functions in the housing market – for example, Reading for cheaper older housing, Slough for flats, Wokingham for larger new housing. On the other hand, the geographical extent of the labour market varies for different categories of worker – more mobile, high earning professionals are capable of purchasing anywhere in Berkshire; the less affluent and less mobile are restricted to specific (geographical) housing markets. It is therefore very difficult to conceive of a direct uncomplicated relationship between the two markets.

Generally, therefore, the evidence from Berkshire adds weight to Hamnett's (1984) contention that the two markets develop in relation to each other. Clearly, at an abstract level this is the case: workers need to be housed. But this statement should be qualified in two ways. First, we need to be clear as to exactly what the 'market' is. For example, is the housing and employment situation in Berkshire characterized by strictly market relations of supply and demand? Disposable income (including non-monetary perks) is significant for determining access to housing – particularly in the absence of a significant council stock – and the supply of housing is conditioned largely by the profit motive. In this situation there is, perhaps, a clearer relationship between the housing and labour markets. But we should not overlook the fact that both are 'imperfect', using the terminology of classical economics. That is, they are conditioned by the effects of the political process on the planning and housing provision systems.

Second, the nature of the relationship between the two 'markets' cannot be pre-determined prior to an empirical analysis of specific cases. Change in the one will effect change in the other, but the relationship could be so highly mediated as to make its nature very ambiguous. In Berkshire the political conflicts around the conservation movement are one example of a mediating influence. So while there is clearly some sort of relationship between housing and labour markets, we cannot anticipate a priori the specific effects of the relationship, both on the economy and on the political alliances in any given area.

Notes

1 This chapter is based on Barlow and Savage (1986, 1987). The research was undertaken as part of an ESRC-funded programme on 'Economic Restructuring, Social Change and the Locality' being conducted by members of the Urban and Regional Studies division of the University of Sussex (see Savage *et al.*, 1987).

2 This issue is being explored in detail by a research project involving Colin Harris and Peter Saunders, examining the social implications of owner occupation in three different British towns, including Slough.

References

Barlow J. (1986), 'Economic restructuring and housing provision in Britain: change in the labour market and change in the housing market', *University of Sussex Urban and Regional Studies Working Paper*, no. 54.

Barlow J. (1987), 'The housing crisis and its local dimensions'. *Housing Studies*, 2(1), 28–41.

Barlow J. (1988a), 'The regionalisation of London's housing market', paper presented at the International Conference of Housing Policy and Urban Innovation, Amsterdam, June.

Barlow J. (1988b), 'The politics of land into the 1990s: landowners, farmers and developers in lowland Britain', *Policy and Politics*, 16, 2, pp. 111–22.

Barlow J. and Savage M. (1986), 'The politics of growth: cleavage and conflict in a Tory heartland', *Capital and Class*, 31, pp. 156–82.

Barlow J. and Savage M. (1987), 'Economic restructuring and housing provision in Berkshire: some implications for the study of housing and labour markets', *University of Sussex Urban and Regional Studies Working Paper*, no. 55.

Brown E. (1986), 'Tee-shot from the boss's golf club', the *Observer*, 18 May.

Crompton R. and Jones G. (1984), *White Collar Proletariat. Deskilling and Gender in Clerical Work* (London, Heinemann).

Fleming S. (1984), 'Housebuilders in an area of growth: negotiating the built environment of central Berkshire', *University of Reading Geographical Papers*, no. 84.

Forrest R. and Murie A. (1986), 'Marginalization and subsidized individualism: the sale of council houses and the restructuring of the British welfare state', *International Journal of Urban and Regional Research*, 10(1), pp. 46–66.

Green A. (1985), 'Recent trends in the service sector in Tyne and Wear and Berkshire', University of Newcastle upon Tyne CURDS Discussion Paper, 71.

Hall P., Breheney M., McQuaid R. and Hart D. (1987), *Western Sunrise: The Genesis and Growth of Britain's Major High Tech Corridor* (London, Allen and Unwin).

Hamnett C. (1984), 'The post war restructuring of the British housing and labour markets: a critical comment on Thorns', *Environment and Planning A*, 16, pp. 147–62.

Hamnett C. and Randolph W. (1986), 'The role of labour and housing markets in the production of geographical variations in social stratification', in K. Hoggart and E. Koffmann (eds), *Politics, Geography and Social Stratification* (London, Croom Helm).

Molho I. and Elias P. (1982), 'A study of Regional Trends in the Labour Force Participation of Married Women, 1968–1977', Institute of Employment, University of Warwick.

Morgan K. and Sayer A. (1988), *Microcircuits of Capital: 'Sunrise' Industry and Uneven Development* (Cambridge, Polity Press).

NBS (1985), *Local Area Housing Statistics: Berkshire* (London, Nationwide Building Society).

Parsons D. (1986), 'Housing Constraints in the Mid-Berkshire Labour Market. A Report to the House-Builders' Federation on Employers' Experiences of Housing Constraints to Recruitment in Mid-Berkshire', Institute of Manpower Studies, University of Sussex.

Pauley R. (1986), 'Riding a high-tech wave', *Financial Times* 30 June.

RCB (1983), *Central Berkshire, Land for the Additional 8000 Houses: Strategic Guidelines*, Royal County of Berkshire Planning Department.

RCB (1985a), *Report of the 1984 Survey of Employers*, Royal County of Berkshire Planning Department.

RCB (1985b), *Development Trends in Berkshire, 1984*, Royal County of Berkshire Planning Department.

RCB (1985c), *Review of Berkshire's Structure Plans. Stage 2 Consultation*, Royal County of Berkshire Planning Department.

RCB (1985d), *Planning Commitments for Housing at March 1985*, Royal County of Berkshire Planning Department.

RCB (1986), *Review of Berkshire's Structure Plans. Stage 2 Consultation*, Royal County of Berkshire Planning Department.
RCB (n.d.), *Survey of New Households. Summary Report*, Royal County of Berkshire Planning Department.

Savage M. (1987a), 'Segmentation and fragmentation in the manual labour market: recent trends in a high tech area', mimeo, University of Sussex, Centre for Urban and Regional Studies.
Savage M. (1987b), 'The white collar labour market in Slough: sex segmentation and occupational careers', mimeo, Centre for Urban and Regional Studies, University of Sussex.
Savage M. (1987c), 'Spatial mobility and the professional labour market: a case study of employers from Slough', *University of Sussex Urban and Regional Studies Working Paper*, xx.
Savage M. (1988), 'Trade unionism, sex segregation and the State: the case of "new industries" in the inter war years', *Social History*, 13, 2, pp. 209–30.
Savage M, Dickens P., Fielding A. (1988), 'Some social and political implications of the contemporary fragmentation of the service class', *International Journal of Urban and Regional Research*, vol. 12, 3. pp. 455–75.
Savage M., Barlow J., Duncan S., Saunders P. (1987), '"Locality research": the Sussex programme on "economic restructuring, social change and the locality"', *Quarterly Journal of Social Affairs*, 3(1); pp. 27–51
Saxenian A. (1985), 'Silicon Valley and Route 128: regional prototypes or historic exceptions', in M. Castells (ed.), *High Technology, Space and Society* (Beverly Hills, Sage).

Witt S. and Fleming S. (1984), 'Planning councillors in an area of growth: little power but all the blame?', *University of Reading Geographical Papers*, 85.

10 *Replacing a surplus population: the employment and housing policies of the London Docklands Development Corporation*

MARK GOODWIN

Introduction

The severe economic decline of Britain since the mid-1970s, and attendant political responses, have resulted in dramatic changes to the geography of the country. As the chapters in this collection make clear, changes in the structure of housing and employment markets have played a key role in this process of uneven development. To a certain extent this has reasserted the 'regional problem', as the social and economic differences between the older industrial regions and the rest of the country have widened. But we have also witnessed a growing diversity within any one region; suburban affluence can increasingly be found adjacent to industrial decay, and the decline of large cities has often been accompanied by the growth of the smaller towns and commuter villages which surround them. Policy changes have reflected this growing local, as well as regional, diversity. Britain has seen a decline in regional policy, and the growth of specifically local policies aimed at the regeneration of particular cities, or even specific parts of these cities (see Duncan and Goodwin, 1988, for a review of 'local' economic policies).

This chapter addresses this change in policy by examining one of these local initiatives, the London Docklands Development Corporation (LDDC), which was set up by the Conservative government in 1981. It was charged with the task of regenerating the eight and-a-half square miles of abandoned docks, wharfs and warehouses that make up London Docklands (Fig. 10.1). The area straddles the River Thames immediately downriver from London Bridge, and when the last upstream docks closed, also in 1981, almost half of its 5,000 acres were derelict or vacant, offering enormous development potential. Since then the LDDC has initiated a large transformation of the area, with housing and employment change at the forefront. Indeed, it is doubtful if any area of similar size in the country has witnessed comparable developments. Dock employment has fallen from 23,000 in 1967 to virtually nil twenty years later. and service sector employment now stands at 70 per cent in what was once a vast complex of port, engineering and refining industry (Goodwin, 1986, p. 20). The number of households in the LDDC area is set to double inside a decade, during which time owner occupation will have risen from 5 to 50 per cent (Coupland and Tuckley, 1985, p. 20; Docklands Forum, 1987, p. 16). Physical and economic change is thus being accompanied by the rapid, and somewhat uneasy, grafting of a new community onto that of London's East End.

But the scale of change is only part of the reason for studying London's Docklands. Equally important is the fact that these recent developments have been guided and promoted by an entirely new state agency committed to restructuring both local employment and housing. In the past, urban regeneration initiatives have usually concentrated on reviving housing or employment, and have rarely sought to combine the two. The LDDC, however, has made no secret of the fact that it seeks not just economic development, but also changes in the local community itself. This will involve a decline in the amount of council housing from 85 to 50 per cent, coupled with a tenfold increase in owner occupation. The LDDC thus provides us with a good opportunity of studying the ways in which housing and employment markets can be linked, or indeed not linked, through policy change at the local level. We will examine the nature of these linkages later in the chapter, by looking in detail at the economic and social changes which have taken place in Docklands. But before we do so, we will set such changes into a wider context by providing an analytical framework through which we can understand the recent restructuring of urban environments in both Western Europe and North America.

The restructuring of urban environments

Like any aspect of urban change, the recent development of London Docklands can only be understood against the background of far wider social forces. In the words of Neil Smith,

> A given built environment expresses specific patterns of production and reproduction, consumption and circulation, and

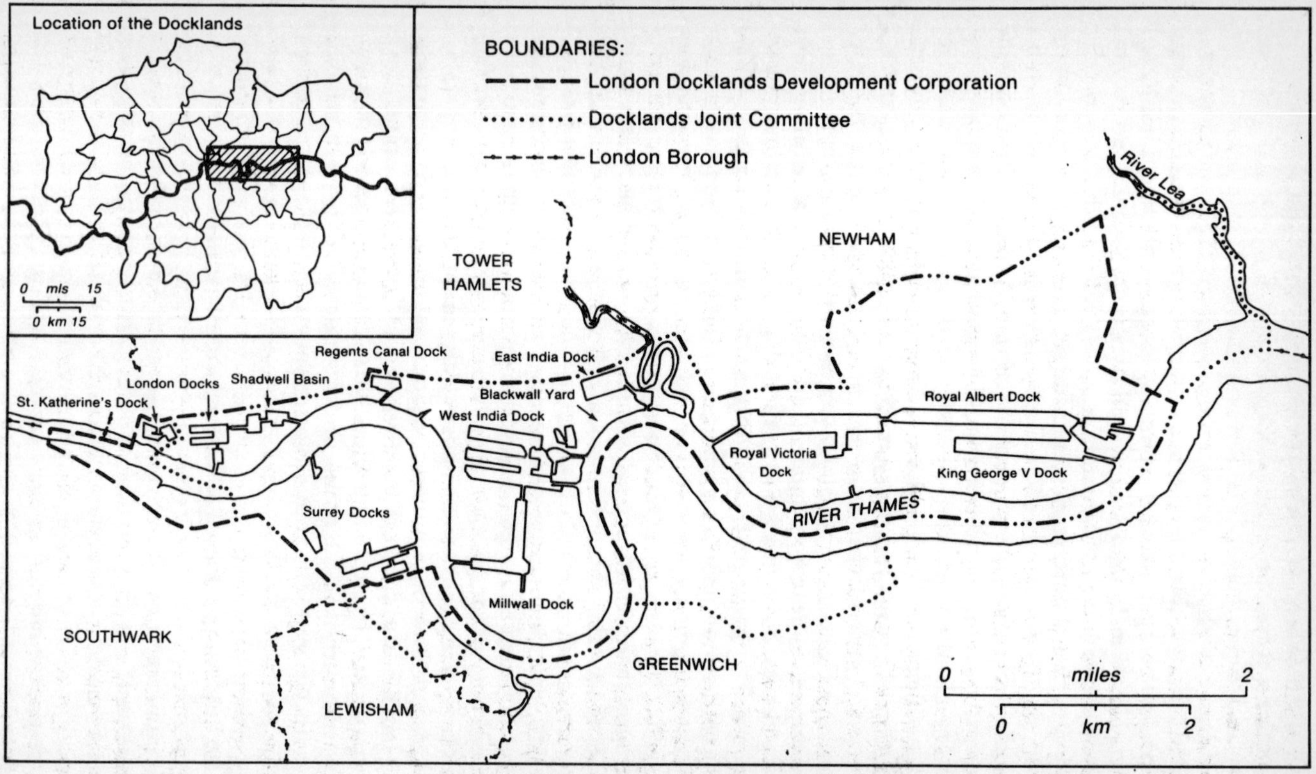

Figure 10.1 Location of London Docklands

as these patterns change, so does the geographical patterning of the built environment. (1986, p. 21)

At present these urban patterns are predicated upon a social system which is continually evolving new social and spatial configurations. This is largely due to the differential nature of investment, again both social and spatial, as capital undergoes its relentless search for a surplus. On the one hand capital is constantly invested in creating those productive, infrastructural and reproductive facilities (factories, roads and houses, for example) necessary to produce surplus value. But, on the other hand, capital is continually withdrawn from created facilities so that it can move elsewhere and so take advantage of social and economic conditions now allowing higher profit rates. The upshot is that created social and physical environments – that is geographies – are continually developed, changed and abandoned (see Harvey, 1982, 1985; Smith, 1984 on the geographical implications of capitalist accumulation).

Often these environments will be formed into particular territories through what Harvey (1985, p. 146) calls the 'structured coherence' of production and consumption. Processes and infrastructures, both social and physical, combine to produce a particular space currently suitable for a specific route in producing surplus value. Forms and technologies of production, inter-industry linkages, patterns of labour supply and demand, physical and social infrastructures, patterns of consumption, all these interact to produce such coherence. Hence, Harvey argues, the emergence of distinctive industrial regions, with distinctive local labour markets, within which labour power with particular specialisms can be substituted on a daily basis. The London Docks were once the centre of one such region of 'structured coherence'.

We should not imagine that people are left as passive agents as 'capital' creates and destroys geographies around them. For it is, of course, people that create and reproduce these social and economic processes, even if they do not do so in freedom from others, nor in conditions of their own choosing. People are able to monitor and learn from their experiences, they adapt, and build economic and political organizations to help further their particular interests. Certain regional and local cultures are also developed, formed by social relations and practices outside capital's logic. Together, these movements and cultures can be important in helping to sustain or destroy the structured coherence of a particular place, but as we shall see, state agencies such as the LDDC are perhaps more powerful in the dismantling and regeneration of particular environments.

More important, however, this 'structured coherence' is often cemented, and reinforced, through the creation of fixed and relatively immobile spatial configurations. But these in turn can act as a barrier to further accumulation. In Harvey's words (1985, p. 150),

> Capitalist development must negotiate a knife-edge between preserving the values of past commitments made at a particular place and time, or devaluing them to open up fresh room for accumulation. Capitalism perpetually strives, therefore, to create a social and physical landscape in its own image and requisite to its own needs at a particular point in time, only just as certainly to undermine, disrupt and even destroy that landscape at a later point in time.

London's docks are a prime example of this 'restless formation and reformation of geographical landscapes' (Harvey, 1985). They represent the creation of an interactive, complementary and spatially fixed assemblage of productive and reproductive structures. In this instance, Victorian capital (and labour) were literally dug into the ground in order to construct a social and physical environment from which an expanded industrial empire of trade and commerce could be developed. The social production of space was necessary in order to overcome space. However, changes in the social relations of production on a world scale leading to the decline of this empire, and global changes in the spatial division of labour, have led to the redundancy of this particular 'spatial fix'. Geographical configurations of production, exchange and consumption have changed, and the production of urban space is taking a new form.

In both Western Europe and North America alike, this has resulted in the dramatic deindustrialization, and reconstruction, of whole urban areas. This is evident in London's Docklands, whose port-related local economy has been completely destroyed by changes in transhipment technology and investment. This has meant the gradual obsolescence (i.e. devalorization) of industrial buildings and infrastructure, so 'freeing' 2,000 acres of land for redevelopment. But obsolete buildings and disused land are of themselves no guarantee of development. Demand must be available in order for capital to be profitably reinvested, and once more the clue to the realization of this process lies in social and economic changes at the global level. These have led to the increasing development of a service-oriented urban economy in the Western world, and a tertiary sector which makes ever increasing demands upon the newly vacant urban space for commercial and administrative functions. This latter trend is itself predicated on the development of a distinct spatial division of labour within the service sector. Routine office tasks have been decentralized to the urban periphery, or even to the regional periphery, while financial, commercial and administrative headquarters have remained within the city centre. Thus certain white-collar activities are decentralized while others are centralized, resulting in an increasing managerial and professional character to the inner urban core and an increasing demand for new high-tech office space in the centre of our cities (see Rees and Lambert, 1985; Smith, 1986).

Given the nature of its economy, London has been affected by this process more than most cities. The recent de-regulation of the City of London has only added to the demand for modern office accommodation. In the world of finance, according to a City property agent,

> The need is for watertight, multi-storey, clear-spanning, regular-shaped factories capable of multi-occupation . . . and maximum internal flexibility. (The *Guardian*, 26 November 1985)

For global finance, then, buildings are seen as 'factories', as just another piece of infrastructure. That the City is unable to accommodate much of this particular infrastructural innovation has not been lost on the LDDC. As their 1985 Annual Report stated

> The City is entering a major phase of evolution and expansion which coincides with Docklands' ability to offer uniquely appropriate space for that expansion. Deregulation will be accompanied by a massive influx of technology as banks, insurance companies and overseas firms enter the London securities market. This is the major opportunity – both for the City and for Docklands. (Carr and Weir, 1986, p. 8)

The LDDC clearly appreciates the links between Docklands' redevelopment and global changes in the world economy. It is important that we do the same. The nature of the latest world recession, and consequent changes in the international division of labour, make it necessary to view London's urban change together with that taking place in other cities. Thus, massive new commercial waterfront developments have taken place in the vacant dock areas of, for example, Chicago, Vancouver, Baltimore, Bristol, New York and Toronto. Other inner-urban areas have seen redevelopment consequent upon changing land uses – ranging from the transformation of Covent Garden in London itself to Les Halles in Paris, the Renaissance Centre in Detroit and Bunkers Hill in Los Angeles. Crucially, these cities have been able to take advantage of social, as well as economic, restructuring. The rise of the 'new service class' of young urban professionals has created a new market for inner-city housing, consumption facilities and leisure provision which the reconstruction of these cities has met. Gradually one form of structured coherence is being replaced by another, and the urban landscape, both social and physical, reflects and reinforces such change.

But more importantly, some cities, such as Liverpool, have been bypassed completely by the new international division of labour. This is a reminder that we must relate these general processes to the specificities of local social relations and local structures. General processes, such as the uneven development of capital accumulation, are commonly seen as non-spatial, as if they float above the real world in some space-less realm. This is not the case. Rather they are always constituted from the start in countless individual places, they are only general in the sense that their constitution is not confined to a single place. Moreover, this means that their particular manifestation in a particular place will always be affected by existing social and physical structures, which will influence to what extent, and in which ways, these general processes work (see Duncan and Goodwin, 1988). We will now sketch out the major ways in which this crucial combination, between the general and the particular, has been realized in Docklands, beginning with a review of the LDDC itself.

The LDDC and docklands development

The London Docklands Development Corporation was designated in 1981, the same year in which the Royal Docks were closed. The former event represented the beginning of a new era in Docklands' development, the latter marked the final act in the previous phase. The closure of the Royals was only the most recent and visible sign of a decline which had been ongoing for over a century. In the mid-nineteenth century, London was the busiest port in the world, by value and volume of trade (Hardy, 1983, p. 6). The initial development of the docks was linked to an expanding empire and the growth of an imperial trade. By the beginning of the nineteenth century the first enclosed docks had been established, and around them grew a vast complex of port-related industrial activity. But this supremacy was not to last, and new phases of international capitalist trade at the beginning of the twentieth century heralded an initial decline. This was hastened by new developments in both production and distribution. Dock activity began to shift elsewhere in Britain, and to northern European ports such as Antwerp, Hamburg, Rotterdam and Amsterdam, all transformed by massive investment to meet these changes.

What little investment London did receive went downstream to the new docks at Tilbury. The old upstream docks, locked into their four peninsulas of Wapping, Rotherhithe, Isle of Dogs and Silvertown, and surrounded by high walls, were run down in favour of their downstream rival, especially with the coming of containerization. The large shipping companies introduced this new 'roll-on, roll-off' technology in the 1960s in an effort to save on labour costs by reducing turn-around time. Land for container storage, and good road/rail links became important. Moreover, shipping companies sought movement to reduce their costs in another way; by employing cheaper non-unionized labour rather than registered dockers. Only one registered docker was required to load a cargo where ten would have been needed before, since containers were loaded and unloaded beyond port confines. This run down of the upstream London docks thus helped to break the hold of organized dock labour, and so cheapen the shipping process.

An overcapacity in both upstream dockspace and warehousing had in fact gradually become apparent since World War II. The war itself had contributed to Docklands' dereliction, as many East End boroughs lost up to 20 per cent of their buildings as a result of bombing. The process was completed by the movement of manufacturing and distribution activities to new suburban estates and greenfield sites. The serious decline of London Docklands soon prompted planning concern and a steady stream of reports and plans emerged from a succession of agencies during the 1970s. These encouraged the notion of the whole upstream docks as one complete planning and development entity – a notion recognized by the setting up of the Docklands Joint Committee (DJC) on 1 January 1974, by the Conservative Government. The DJC was a single agency, and, as the name implies, was a joint committee composed of representatives from the five local authorities within its boundaries and from the Greater London Council (GLC). It could also co-opt additional members, usually drawn from the Port of London Authority, local trades unions and local community groups (see Figure 10.1, p. 256).

The DJC was charged to draw up and implement a Docklands-wide regeneration strategy. As stated in their London Docklands Strategic Plan, published in 1976, the committee wished to improve, rather than replace, existing housing and employment activities, and to aim these improvements at local people. The major components of this strategy were to be a large public housing programme, additional maufacturing industry and improved transportation links. But in order to achieve this the committee was given no powers or resources over and above those of its constituent local authorities (Greater London Council, Tower Hamlets, Newham,

Greenwich, Lewisham and Southwark). The local authorities and the GLC delegated their development control functions to the DJC, but, apart from this, it was not an executive agency. Its role was primarily an advisory one of co-ordination and planning.

At first it proved successful, and after producing a whole range of discussion papers for local comment it published a draft 'London Docklands Strategic Plan' in April 1976. Although not a statutory plan, the LDSP sought to guide development into its broad framework over the next twenty-one years. This proved the high point – for one year, 1976 to 1977, Labour controlled all the government agencies involved, and co-ordination was relatively easy. But just as implementation of the plan began in the late 1970s, the DJC ran foul of local and national expenditure cuts. In addition, political factors also conspired against its success. Gradually the Conservatives began to claw back the levers of power. In 1977 they gained the GLC, and in 1979 they returned to Westminster. The five Labour boroughs were left isolated in a committee with hostile partners. The hostility was soon made felt, and the DJC came under increasing criticism for not achieving the goals of its own strategic plan.

Of course, three years' operations out of a proposed twenty-one does not give much time for implementation, especially when no additional powers are given over finance and land. But such protests did not hold much sway with the new central government, which promptly stepped in to impose an appointed body onto the people and the Labour boroughs of London's Docklands. This body was the LDDC, and it represented an entirely new innovation in the field of urban regeneration in Britain (see Duncan and Goodwin, 1988, for a review of the LDDC and other Conservative initiatives in the area of urban regeneration).

The new departure was justified by Michael Heseltine, then Secretary of State for the Environment, in the following terms:

> . . . for London's docklands . . . I do not think that the present arrangements can meet the particular problems and opportunities of the area . . . there is a need for a single-minded determination not possible for the local authorities concerned. To meet the challenge before us I am proposing to take general powers to enable me to set up Urban Development Corporations. (DoE Press Release, 14 September 1979)

These 'general powers' proved extremely wide. They were granted in the 1980 Local Government, Planning and Land Act, which enabled the Secretary of State to set up an Urban Development Corporation if he or she considers it 'expedient in the national interest to do so'. The Act gave the new Corporations powers to:

1 Acquire, hold, manage, reclaim and dispose of land and other property;
2 Carry out building and other operations;
3 Seek to ensure the provision of water, electricity, gas, sewerage and other services;
4 Carry out any business or undertaking for the purpose of the object; and
5 Generally do anything necessary or expedient for the purpose of the object, or for purposes incidental to those purposes.

These powers, of course, overrode those of the elected local authorities, all Labour controlled in 1981. In effect, local powers of development control were taken out of the electoral orbit of local government and placed in the hands of a non-elected body answerable only to the Secretary of State.

Two UDCs were set up following the Act, the LDDC in London and another in Merseyside. Four more followed in 1986, in Manchester, Tyne and Wear, Teesside and the West Midlands, one in Cardiff was announced in 1987, and one in Sheffield in 1988. Several 'mini-UDCs' are also promised. Although the government claimed their sweeping powers were modelled on those of the New Town Development Corporations, they are in fact similar in form only. In content they are quite different. New Towns represent the physical embodiment of Fabian socialism and state planning under the terms of a social-democratic compromise between capital and labour (which is presumably why their programme was halted at every opportunity by successive Conservative governments). In contrast Urban Development Corporations represent a local executive agency of a central government which wishes to reject any compromise with labour, and the one in London, by its own admission, is run according to the criteria of developers and financiers. Commenting on the lack of new employment created by the LDDC in its Enterprise Zone, Nigel Broackes, its first chairman, told the House of Commons Environment Committee that, 'It is not a brilliant record in creating jobs', but it is 'a pretty good record in filling up the Isle of Dogs which is our first priority' (8 August 1983).

Regeneration is based upon encouraging private sector investment through the preparation and marketing of land and infrastructure, and lessening the influence of public planning. As its Chief Executive, Reg Ward, says;

> We have no land use plan or grand design; our plans are essential marketing images. The feeling of cohesion in docklands comes in retrospect. (*The Times*, 18 November 1986)

The removal of development power from locally elected authorities to a centrally appointed one, is no accident. Nigel Broackes saw the LDDC as capital's answer to local electoral politics; 'Why face aggravation from Councils opposed to the profit motive and home ownership?' (the *Guardian*, 3 July 1981). Why indeed, when you have a government prepared to take development control away from elected local authorities?

The government justifies these moves not only through references to the inefficiency of local democracy but by also referring to the 'national interest', and to the need to ensure that such sums of money are 'properly spent'. These justifications were often heard during the House of Lords Enquiry into the setting up of the LDDC – itself ironically only held because the government in its haste put forward a hybrid order when designating the corporation. This gave a chance for local authorities and local community groups to oppose the LDDC, but although the Lords Committee, which heard the enquiry, had 'no doubt that the opposition to the proposed transfer of development control . . . is perfectly genuine' and were 'very conscious that to transfer development control over so wide an area from democratically elected councils to a body appointed by the Secretary of State is a step which is not easily to be justified' (Report of House of Lords Select Committee, paras. 8.3; 8.8), they nevertheless concluded that the possibility of quick commercial development should be given more weight than the exercise of local democracy.

The government told the enquiry on the first day that it had to ensure 'that the national interest is being met' (House of Lords Enquiry, Day 1, p. 48). Whatever this interest is, it is not to be reached through using the existing channels of local government, which 'tend to look too much to the past and too exclusively to the aspirations of the existing population' (Report of House of Lords Select Committee, para. 6.4). Instead the single-minded determination alluded to by Heseltine, can only be guaranteed through 'direct government control through a government nominated body' (House of Lords Enquiry, Day 15, p. 18). When coupled with a budget of £60–80 million per annum, it is only to be expected that this single-minded determination has had considerable effects on both the local labour and housing markets.

Changes in Docklands' labour and housing markets

At first sight it appears that the strategy of the LDDC is based on a close linking of local labour and housing markets. Indeed, the House of Lords Enquiry was told of 'a strong link between housing and economic/industrial potential' (Day 3, p. 13). The LDDC developed this theme in subsequent statements. Their first Corporate Plan (never made publicly available in its full form) emphasizes that two of the main problems to be overcome in 'The Regeneration Task' are first, too many unskilled workers and old people and second the limitations of existing labour skills, geared to traditional industries (which) are unlikely in the short term to adapt to the needs of potential new industries' (LDDC, 1982, pp. 4–5).

The LDDC's initial Information and Research Officer stressed the same point when he told a conference (again not open to the public) that the problem was one of 'a surplus capacity of people' (16 October 1982). He later divulged that the LDDC has plans to 'increase the population of the area', indicating that the surplus is not one of an absolute nature – rather it is a surplus of the wrong type of people for the specific type of regeneration the LDDC has in mind. This was confirmed by the LDDC's chairman, who, in his introduction to the Corporate Plan, wrote that: 'Housing was identified as our first priority and our achievements have been highly significant in taking the first steps in repopulating Docklands.'

Presumably the need for 'repopulation' was somewhat lost on the 40,000 people who already lived in Docklands. Maybe for their benefit, the Corporate Plan becomes more explicit. It states that one of the priorities on which the LDDC's housing policies are based is the need 'to provide the variety, numbers and quality of housing needed to re-establish in Docklands a population with the youth and skills required for industrial and commercial expansion there' (p. 17). Thus a changed housing market would be linked to a changed labour market, through the removal of a 'surplus population' followed by a subsequent 'repopulation'. In turn this would underlay 'industrial and commercial expansion'. It is undoubtedly true that both labour and housing markets have undergone major transformations during the LDDC's time of operation. But a closer examination of the nature of these changes reveals that the links between them are not quite of the type propounded by the LDDC.

In terms of housing the LDDC has deeply affected both the volume and character of the local housing market, although it is not a statutory housing authority and has no responsibility in the field. Its influence arises from its role in 'stimulating and facilitating development by private developers' (Corporate Plan, p. 16), mainly through assembling, clearing and allocating sites. At the time of designation in 1981, the LDDC area contained 14,881 households, only 5 per cent of whom were owner occupiers (Coupland and Tuckley, 1985, p. 20). Eighty-two per cent were rented from the local authorities, this dominance of public sector dwellings representing the legacy of postwar local authority redevelopment following slum clearance and bomb damage. This was matched by a dominance of the public sector in terms of disposable land, the majority of which was owned by the local authorities and public utilities including the Port of London Authority.

Crucially, the LDDC has been able to take over this publicly owned land free of charge through its vesting powers. Using a vesting order issued by the Secretary of State, land can simply be transferred from the local authorities to the LDDC. In this way it has been able to quickly build up a very sizeable land bank, which was denied completely to the DJC. Through controlling land, often bought and prepared by the boroughs, the LDDC has been able to dramatically change the housing market. By September 1985 there had been 7,350 new dwellings started in the LDDC area, a 50 per cent increase in the housing stock in four years. Sites were then available for a further 6,600, representing a total housing potential roughly equal to the 1981 stock. Of those started, 82 per cent were for sale, 14 per cent for rent and 4 per cent for shared ownership (Docklands Forum, 1987, p. 16). This is despite an assurance given to the House of Lords Select Committee by the LDDC Chairman, Nigel Broackes, that the tenure mix would be 50–25–25 respectively. If these trends continue, owner occupation would increase from 5 to 50 per cent – creating a vast change, not only in housing tenure, but also in the size and character of the local community (Docklands Consultative Committee, 1985, p. 12).

This is because local people have largely been excluded from the property bonanza that has suddenly erupted around them. The new speculative housing market in Docklands has been largely realized through the purchasing power of young, two-income households from outside the area. In Newham, where the majority of low-cost homes have been built, a council survey of 800 new properties found that only 4 per cent were occupied by ex-tenants of Newham Council. The LDDC has reduced its own figure for local purchasers in Newham from 60 per cent to 12 per cent. As more and more houses are built at more expensive prices, fewer and fewer local people can afford to buy them (Docklands Consultative Committee, 1985, p.13; Docklands Forum, 1987, p. 17). Indeed, evidence from surveys of local incomes suggests that 'There is little scope for further owner occupation among the existing Dockland boroughs population . . . the vast majority . . . have not got the income to enter into owner occupation' (Docklands Forum, 1987, pp. 22–3).

It is certainly true that the policies of the LDDC to promote private housing development have failed to meet local housing needs. In this sense, despite 7,000 additions to the stock, the local housing market has deteriorated badly. The three Docklands' boroughs accepted 1,682 families as homeless in 1979/80. By 1985/86 the number of families in priority need had risen to 3,059, an increase of 81 per cent. Over the same period the rest of inner London's homelessness increased by 38 per cent. Meanwhile, outside priority need, a further 30,000 applicants were seeking housing through the waiting lists of the three Docklands' boroughs – Newham, Southwark and Tower Hamlets (Docklands Forum, 1987, pp. 2–8). However, meeting local housing need has never been seen by the LDDC as one of its roles, and in its own words it prefers to 'complement the efforts of the Housing Authorities in the area, whose primary objective is to respond to housing needs' (Corporate Plan, p. 16). But by immediately vesting 600 acres of public land which the boroughs had prepared for badly needed housing sites, the LDDC has actually hampered, rather than complemented, this effort. The Local Authorities now lack money, because of public expenditure cuts, and land, because of the vesting powers of the LDDC, with which to meet housing need.

In place of local authority housing to meet local need has come a new, overwhelmingly private and very expensive, housing market. Docklands quickly became established as perhaps the prime site in a booming London residential property sector. The marketing schemes of the LDDC and the developers have succeeded in adding social desirability to the intrinsic benefits of a location adjacent to the City of London. Until the 1989/90 property slump penthouse flats overlooking the river regularly sold for over £500,000, the top end of a luxury market in which, according to Chestertons estate agents, 'The scope for capital appreciation and potential growth is without comparison' (*The Times*, 18 November 1986). Land speculation has only worsened, rather than helped, local housing need. Perhaps, though, this is no accident. In the words of the Corporation's Chief Executive, comparing its policy to that of its predecessor: 'Housing is part of our regeneration task whereas the Docklands Joint Committee looked at housing solely in terms of need' (Ambrose, 1986, p. 225).

The other 'part' of the 'regeneration task' of the LDDC, and the second half of its plans to restructure the society and economy of Docklands, is centred around the economic development of the area. Claims and counterclaims abound regarding the precise nature of changes in the labour market since the LDDC took over development powers, especially concerning the extent of jobs attracted to the area by the boroughs before 1981, but arriving afterwards. Table 10.1 presents an outline of total employment change in Docklands from 1981 to 1985.

Table 10.9 Total employment change in Docklands 1981-5

Occupation	Employment in 1981 Number	Employment in 1981 Per cent	Employment in 1985 Number	Employment in 1985 Per cent	Change 1981-5 Number	Change 1981-5 Per cent
Agriculture	0	0	33	0	+33	N/A
Energy	645	2	480	2	−165	−26
Mineral Extraction	1,368	5	392	1	−976	−71
Metal Goods	1,130	4	1,931	7	+801	+71
Other Manufacture	7,901	29	3,871	14	−4,030	−51
Construction	1,964	7	1,603	6	−361	−18
Distribution	4,220	16	9,001	32	+4,781	+113
Transport	5,652	21	3,031	11	−2,621	−46
Banking	1,452	5	3,498	12	+2,046	+141
Other Services	2,851	10	4,340	15	+1,489	+52
Other	30	0	0	0	−30	−100
Total Employment	27,213	99	28,180	100	+967	

Source: Church, 1986, p.27.

Thus, although the LDDC in its 1984–5 Annual Report claimed the addition of 5,700 jobs to the Docklands' economy, the actual net gain was less than a thousand because of closures, movements and redundancies. By 1987, the LDDC claimed that 8,000 jobs had been created in Docklands, but continued job loss has restricted the overall gain. The LDDC has actually stimulated some of this loss, either directly through redeveloping land, or indirectly through increasing land prices and rents. Moreover, many of the incoming jobs, especially in the larger newspaper plants and financial offices, are in fact redistributed from elsewhere in London and cannot be claimed as 'new' employment.

The LDDC seems to have made little impact in terms of job numbers – although it claims that projected developments on Canary Wharf and in the Royal Docks will add up to 100,000 jobs. The proposed development at Canary Wharf on the Isle of Dogs, represents the largest single office development ever undertaken in Britain and includes three 850-foot towers – the highest in the country. The scheme, described in its publicity as 'the new financial centre of Europe' hopes to meet the need for new office space in the City of London following the de-regulation of the Stock Exchange. Despite financial difficulties, the developers claim they are still on course to complete the 10 million square feet development and building is well under way. But the employment impact is uncertain. Consultants' reports differ both in the amount of jobs to be generated, and in the percentage of these that will go to local people (see Church, 1986, pp. 14–16).

Although the future is uncertain, what can be said at present is that the LDDC has helped to significantly alter the nature, if not the number, of Docklands' jobs. Manufacturing, which in 1981 accounted for 41 per cent of Docklands' employment, had declined to 24 per cent by 1985. The service sector had increased from 52 per cent to 70 per cent. The changed structure of the local economy has had major effects on the local labour market. It has meant the widening of that market, as the new Docklands' jobs have primarily been taken by those living elsewhere. In its industrial prime, the area could claim a relatively closed labour market. Access to employment in the docks, especially, was jealously controlled through local social networks – both informal and trade union based. A 1976 survey of the Isle of Dogs found that 87.5 per cent of economically active adults were in manual employment – compared to 51 per cent for Greater London as a whole (Cole, 1983, p. 275). But the industrial foundation for this type of employment has now been lost, and the area's new service sector economy draws its employees from a very wide regional, rather than a local, labour market. Other social networks are now beginning to operate, this time based around the City rather than the docks. In spite of new developments, Docklands' unemployment has risen almost 50 per cent from 1981 to 1986 and male unemployment stood at 32 per cent – one in three of the local male population, and female unemployment has doubled from 8 to 16 per cent (Church, 1986, p. 12).

The Enterprise Zone on the Isle of Dogs has been the principal focus of the LDDC's attempt to regenerate the economy of Docklands. A monitoring study by a consultancy firm appointed by the government concluded:

> . . . the EZ's effect on bringing employment into the local area is not matched by a corresponding increase in job opportunities for local people. (Docklands Consultative Committee, 1985, p. 22)

A similar survey carried out by the Association of Island Communities found that of the 1,400 new jobs in the Enterprise Zone by 1985, just twenty-eight were held by residents of the Isle of Dogs. The LDDC itself estimates that 28 per cent of Enterprise Zone employment is held by residents of the three Docklands' boroughs – including those who have moved in to the new housing. Another survey of new employment in the Newham section of the LDDC area found that only 19 per cent of these jobs were taken by residents of the local borough (Church, 1986, p. 13).

Thus, most new jobs in Docklands conform to the usual London pattern of a very wide labour market (both spatially and socially), especially in office, technical and managerial employment. One would not necessarily expect local jobs to go to local people in these types of occupations, but Docklands' residents have had to watch their economy, and their labour market, shift to this conformity over the past few years. The nature of their reaction, if this continues, remains to be seen. As *The Times* pointed out, chief among Docklands' most 'persistent and as yet unresolved problems . . . is the very real prospect that local people, whether or not they are elbowed out in the property boom, seem unlikely to share significantly in any future jobs bonanza, except perhaps on the menial periphery' (18 December 1986).

The people of Docklands, whose manual skills once provided the foundation for the hub of a trading empire, now seem destined to enter the local labour market mainly in the service sector. A confidential consultants' report commissioned by the LDDC on Canary Wharf, but leaked to the press, estimated that only 1,800 local people would gain jobs in the new development (out of between 40,000 and 50,000) mainly in cleaning and catering (Church, 1986, p. 15). As the *Guardian* sarcastically put it, 'Twenty-eight jobs on Canary Wharf for local unemployed youngsters is regrettably no substitute for 40,000 wheeler dealers imported from Threadneedle Street' (7 October 1985). These new jobs are now firmly part of a quite different labour market, which although restricted in parts, is dispersed socially and geographically.

The new Docklands' residents, who reside in the luxury flats and speculative housing, are also part of this dispersed London-wide labour market. Although reliable data is difficult to find, the indications are that the vast majority of new residents work outside the Docklands area. An early survey of new residents in Newham found that two-thirds worked outside Docklands, and of those who worked locally, 64 per cent intended to move home anyway. It would seem to be that the case as put forward by the LDDC is misleading, concerning the links between the labour and housing markets – the latter does not supply the former with 'the youth and skills required for industrial and commercial expansion' (LDDC, 1982, p. 17). The labour market draws its supply from a far wider area, and those who have entered the Docklands' housing market overwhelmingly work elsewhere. Close links between the two markets can, however, still be found – but they must be sought in a different, and somewhat hidden, form.

The LDDC and the hidden links between Docklands' labour and housing markets

Whatever the nature of housing and employment change in London Docklands, and the position of local people, it is impossible to deny the extent of the transformation which the LDDC has initiated. The Corporation itself views this in the following manner:

> The original construction of these docks in the 19th century provides the precedent for such an audacious expansion. Now, as then, they are an imaginative response to technological progress, a symbol of the vitality and enterprise of our age. (Carr and Weir, 1986, p. 7)

To understand them, however, these developments must be seen as more than just an 'imaginative response' and a 'symbol of vitality and enterprise'. The development of the docks, and their subsequent closure, must be seen as just one part of the economic and social restructuring of British capitalism. As we noted earlier, the Docklands' environment was created in order to take advantage of an expanding trading empire. When international changes, in both economy and society, led to its decline this particular built environment had outlived its usefulness. The specific type of spatial coherence that Docklands represented has in fact become as redundant as the people who once worked there. What we are now witnessing is the rise of another distinct spatially coherent landscape, but this time built on the service sector rather than on manufacturing and shipping. As in many other 'favoured cities', land is being revalorized through the switch to a service sector economy, with a special emphasis on corporate and financial control functions, and the revival of an inner-urban residential market.

London Docklands has proved no exception to these general processes, but they have been mediated through particular sets of local relations. The LDDC has been vital to this mediation in two main ways. First, it represents the key to a changed political structure in the area. The local authorities have had planning and development powers taken from them and handed to this unelected organization. This process is one part of the Conservative Government's aim to replace the public sector with the private in all walks of life. Thus, the Docklands Joint Committee, a body dominated by local authority representatives, wanted to regenerate the area through public housing and manufacturing expansion to meet the needs of local people. The very nature of the LDDC ensures that no opposition will be met in changing this prospectus to private housing and commercial development. In contrasting the Corporation to the boroughs, an LDDC officer commented that 'We have no allegiances, no favours to repay' – which could be interpreted as 'we have no democratic procedures to follow and no electoral promises to keep' (Goodwin, 1986, p. 29). Effectively, an unelected institution is replacing the powers of local government in order to carry out a market-led regeneration of the inner city. As the Planning Officer of one Docklands borough put it, the LDDC 'is designed to shift land and assets from the public to the private sector' (Goodwin, 1986, p. 28).

The success of this shift is dependent upon the ability of the LDDC to promote private investment. But it is vital to realize that this in turn is based on the successful assembling and marketing of land, and that this has used large sums of public money. Decline and dereliction on their own are insufficient to guarantee redevelopment. Private development has to be nudged into gear through public subsidy. The LDDC has spent 77 per cent of its total budget on land acquisition, reclamation and treatment (Church, 1986, p. 5), and in addition has vested land which had benefited from previous public investment. In effect, despite its marketing image as a private sector success story, land in Docklands has in fact been transferred from the public to the private sector by the LDDC acting as a state developer. As Nigel Broackes put it, 'We are . . . a property based organisation offering good value' (*Docklands Action News*, No. 11, January 1983). This good value enables volume house builders to buy land at half the market price (Ambrose, 1986, p. 225), so one Dutch developer was sold a site on the Isle of Dogs for £880,000 after the LDDC had spent £2½ million on purchasing and clearing the land (DCC, p. 13).

The LDDC's approach to regeneration essentially consists of selling Docklands off as quickly as possible. Sites are serviced, prepared and marketed using a public subsidy. Through such subsidy, the LDDC is vital in creating the potential for higher land values, and at the same time ensuring that these are passed on to private capital. Land in Bermondsey which would have sold for £300,000 an acre in 1983, commanded a figure of £3 million only five years later. Other land for office development can cost up to £15 million an acre (the *Daily Telegraph*, 20 August 1987). A key means of facilitating this increase and transfer, is through changing 'the traditional perception of Docklands as a second rate market' (LDDC, 1982, p. 14). As Reg Ward, the Chief Executive of the LDDC put it, 'We needed to get people to see Docklands with different eyes.' He went on to add that attracting large private sector housing developments was crucial to this (*Financial Times*, 17 July 1986). Four years earlier he had stated that owner occupation is 'the spearhead of the redevelopment of Docklands. It is the most important route to changing the perspective of the area' (*Roof*, July/August, 1982, p. 17). This is where developments in the housing and employment fields can be linked. They are twin parts of a recursive duality, in that each in turn affects, and is affected by, the other. In this instance both are seen as vital to enhancing the image, changing the perspective and hence increasing the land values of the area.

Thus changes in the housing market and in the local economy are the twin spearheads of a cumulative process of speculatively led, and financially determined, development. Each needs the other to promote the 'image', 'perception' and value of the area. One without the other would not be sufficient to attract the necessary investment. Twin changes in the labour and housing markets of the area are essential to the revalorization of Docklands. The real links can be found in the market place, where land is allocated not according to a public plan, but according to the heavily subsidized private cheque book. In the Corporation's own words, 'Our plans are essential marketing images' (LDDC Chief Executive, *The Times*, 18 November 1986). This image is dependent upon changes in both the labour and housing markets. This is why the surplus population must be replaced and the community rebuilt. Not because the newcomers to the housing market will provide the youth and skills for the labour market (which will come from further afield), but because 'the poor . . . image of the area as the backyard of London keeps land values depressed' (LDDC, 1982, p. 4). All the LDDC is doing is sweeping the backyard. The consequences if you happen to be in front of the broom are rather different in kind to those occurring if you come behind.

But the exact nature of these consequences, for those in front or behind the broom, cannot be predicted. By definition, a market-led strategy cannot provide close links between the labour and housing markets of a local area – especially an area in the centre of such a dispersed labour market as inner London. Land is being sold in Docklands for the best return, almost irrespective of use. The LDDC have opened a Pandora's Box, but have no control over the contents. Instead these are decided by property developers and finance companies. According to one representative, the LDDC was set up 'to short circuit planning procedure' with a brief to 'look at what the market place can offer' (Conference, University of Sussex, 16 October 1982). Those seeking a policy to link labour and housing markets at the local level would be better advised to strengthen, rather than weaken, planning procedure, for only close control over development can ensure appropriate land use, which in turn is necessary for the two markets to be brought closer together.

Acknowledgements

The author would like to thank Simon Duncan, Mike Savage and the editors for helpful comments on an earlier draft, and owes a special thanks to Andrew Church who not only provided helpful discussion but also allowed the author to see the unpublished manuscript of his 1988 article.

References

Ambrose P. (1986), *Whatever Happened to Planning*? (London, Methuen).

Carr M. and Weir S. (1986), 'Sunrise City', *New Socialist*, September.

Church A. (1988), 'Urban regeneration in London Docklands: a five year policy review', *Environment and Planning*, C, 6, pp. 187–208.

Church A. and Hall J. (1986), 'Discovery of Docklands' *Geographical Magazine*, December.

Cole, T. (1983), 'Life and labour in the Isle of Dogs', PhD thesis, University of Oklahoma.

Coupland A. and Tuckley W. (1985), 'Docklands for Sale', *Roof*, November/December.

Docklands Consultative Committee (1985), *Four Year Review of the LDDC.*

Docklands Forum (1987), *Housing in Docklands* (London, Docklands Forum).

Duncan, S. and Goodwin, M. (1988), *The Local State and Uneven Development* (Cambridge, Polity Press).

Goodwin M. (1986), 'Locality and local state; economic policy for London's Docklands', *Sussex Working Paper in Urban and Regional Studies*, 53.

Hardy D. (1983), 'Making sense of London's Docklands; processes of change', *Geography and Planning Paper, 9*, Middlesex Polytechnic.

Harvey D. (1982), *The Limits to Capital* (Oxford, Blackwell).

Harvey D. (1985), 'The geo-politics of capitalism' in D. Gregory and J. Urry (eds), *Social Relations and Spatial Structures* (London, Macmillan).

Joint Docklands Action Group (1982), *Private Housing – The Key to Docklands Regeneration* (London, Joint Docklands Action Group).

London Docklands Development Corporation (1982), *Corporate Plan*.

Rees G. and Lambert J. (1985), *Cities in Crisis* (London, Edward Arnold).

Smith N. (1984), *Uneven Development* (Oxford, Blackwell).

Smith N. (1986), 'Gentrification, the frontier and the restructuring of urban space' in N. Smith and P. Williams (eds), *Gentrification of the City* (London, Allen and Unwin).

Index